COMPUTER GRAPHICS
Techniques and Applications

COMPUTER GRAPHICS
Techniques and Applications

Edited by

R. D. PARSLOW
Department of Computer Science
Brunel University, Uxbridge, England

R. W. PROWSE
Department of Electrical Engineering and Electronics
Brunel University, Uxbridge, England

AND

R. ELLIOT GREEN
Scientific Control Systems Ltd.
Berners Street, London, England

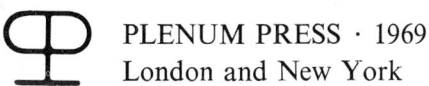

PLENUM PRESS · 1969
London and New York

Plenum Publishing Company Ltd.,
Donington House,
30 Norfolk Street,
London, W.C.2

U.S. Edition published by
Plenum Publishing Corporation,
227 West 17th Street,
New York, New York 10011

Copyright © 1969 by Plenum Publishing Company Ltd.

Reprinted 1970

All Rights Reserved
No part of this book may be reproduced in any form
by photostat, microfilm, or any other means without
written permission from the Publisher.

Library of Congress Catalog Card Number: 68/58992

Printed in Great Britain by
Latimer Trend & Co Ltd, Whitstable.

Contributors

S. BIRD, *Automation Division, The Marconi Company Ltd., Chelmsford, Essex, England.*

P. M. BLACKALL, *CERN, Geneva, Switzerland.*

R. A. CHAMBERS, *Experimental Programming Unit, University of Edinburgh, Edinburgh, Scotland.*

D. R. EVANS, *Royal Radar Establishment, Malvern, England.*

A. E. P. FITZ, *Ministry of Defence (Navy Division), London S.W.6, England.*

C. W. GEAR, *Department of Computer Science, University of Illinois, Urbana, Illinois, U.S.A.*

R. Elliot GREEN, *Scientific Control Systems Ltd., Berners Street, London, England.*

F. M. LARKIN, *UKAEA, Culham Laboratory, Culham, Abingdon, Berkshire, England.*

J. C. LASSALLE, *CERN, Geneva, Switzerland.*

C. MACHOVER, *Information Displays Inc., Mt. Kisco, U.S.A.*

S. M. MATSA, *IBM New York Scientific Center, 410 East 62nd Street, New York, U.S.A.*

D. C. McDOUALL, *Standard Telephone and Cables Ltd., Data Systems Division, Cockfosters, Hertfordshire, England.*

D. MICHIE, *Experimental Programming Unit, University of Edinburgh, Edinburgh, Scotland.*

H. H. POOLE, *Raytheon Company, Wayland, Mass., U.S.A.*

R. W. PROWSE, *Department of Electrical Engineering and Electronics, Brunel University, Uxbridge, England.*

M. A. RUBEN, *Digital Equipment Corporation, Maynard, Mass., U.S.A.*

A. R. RUNDLE, *Elliott Automation Systems Ltd., Borehamwood, Hertfordshire, England.*

F. E. TAYLOR, *National Computing Centre, Manchester, England.*

B. T. TORSON, *Rolls-Royce Ltd., Aero Engine Division, Derby, England.*

C. VANDONI, *CERN, Geneva, Switzerland.*

P. E. WALTER, *County Architect's Department, West Sussex County Council, Chichester, Sussex, England.*

A. YULE, *CERN, Geneva, Switzerland.*

Foreword

About four or five years ago one began to hear about the enormous interest being taken in on-line consoles and displays. Nothing much was done with them, but computer men felt that this was the way computing ought to go: one might dispense with cards, and overcome many of the problems of man–machine communication. It quickly appeared that, as with computers, there had been a great under-estimation of the amount of work involved, of the difficulties of programming, and of the cost. So it began to emerge that graphics was not the ultimate answer, in spite of superb demonstrations where one might watch a square being converted into a cube and then rotated.

But my mind goes back to 1951 and the first computers. There, there were demonstrations of arithmetic speed and storage facility; but not much idea of actual use. However, we now understand how to use computers, and in the last year or two, significant developments in the field of graphics have led to genuine applications, and economic benefits. The equipment is still expensive, but it is becoming cheaper, more uses are being found, and I believe that we are just at the stage when the subject is gaining momentum, to become, like computers, a field of immense importance.

This book, and the symposium at which the papers were first read, will generate ideas for new applications in the minds of those who could use graphics, and further steps will be taken in using the computer as a tool. For it is not only to the specialist, but to all who need the power inherent, but so often locked up, in the computer itself, that graphics is bringing its benefits.

Director,   GORDON BLACK
National Computing Centre,
Manchester,
England.

Preface

"One picture is worth a thousand words", and in computing it is certainly true that one picture can be considerably more valuable than several yards of lineprinter output. This is all the more true if a scientist or a business executive has to interpret the output and take further action on it with the computer.

The graphics terminal opens up a completely new range of fields of application for computers. For the first time an executive or engineer can have direct access to the power of a computer, communicating in visual terms which are natural to man.

The potential recently revealed for interaction between computer and user is vitally important for the greater application of these machines in all spheres of industry, commerce and scientific development. The interactive graphic terminal transfers the computer from a cumbersome specialist "tool" into a "colleague" helping to work out a solution to problems during a dialogue through the common visual link.

To gain the most immediate benefit from computer graphics it is essential for all who are involved to be informed about existing applications and about the trend of further development. These are the computer technologists designing for the future; the computer manufacturers producing for today; the researchers and designers in every field whose problems might be all the more easily solved by the new means available; and the industrial and technical managers who can now begin to think of computers as an accessible means of making their organisations more efficient. Only a combination of the thought and efforts of all these parties can ensure the speediest and most effective development of the new techniques and equipment.

For this reason the Computer Science Department, Brunel University, decided to organise an International Computer Graphics Symposium, where carefully selected themes were covered by foremost authorities from the USA and UK.

A glance at the Table of Contents will show how contributors were invited from the world's leading academic institutions, manufacturing firms, research establishments and industrial and commercial users.

The material for the Symposium (held at Brunel University, Uxbridge, England, in July 1968) was kindly made available and specially edited with relevant additions and amendments for this book. The aim has been to allow those people employed in the field to learn of each other's activities and for those who can benefit from their efforts to discover what facilities are being made available.

The book is in four parts.

PART 1 covers the systems, equipment and software, which can now be employed; the general stage of development in the USA and

UK; and the trends for the future. It serves as an introduction to the field for non-experts and also as a valuable résumé for the initiate.

PART 2 consists of specific applications in science and industry, with several case histories of successful installations. These cover many fields, from architectural design and costing to nuclear physics, aircraft engineering and stock control.

PART 3 is for the computer technologist and is a review of material which was presented and discussed at a Specialist Session which followed the Symposium.

PART 4 is devoted to computer graphics hardware, which is presently available. It includes manufacturer's descriptions of a wide variety of equipment.

A glossary has been provided to explain graphics terms used in the book, so that all interested readers can obtain maximum value from the ideas expounded, unhindered by unfamiliar terminology.

This book has been designed to cover the field of Computer Graphics in a logical and comprehensive manner. It can be read as a whole to review all the important aspects of the subject or studied piecemeal as a report on particular topics.

Acknowledgements

Special thanks are due to:

>   the authors, for adapting their papers for publication;
>
>   Sydney Paulden, who acted as publication adviser to the editors;
>
>   Rita Gregory, for her administrative contribution;
>
>   and to the many willing helpers who devoted so much time and energy to the organisation of the Brunel Symposium and the production of this book.

ROBERT D. PARSLOW
ROGER W. PROWSE
RICHARD ELLIOT GREEN

October 1968

Contents

Contributors	v
Foreword	vii
Preface	ix
List of illustrations	xiii

PART 1
Systems, Equipment, Techniques, and Trends

What has Computer Graphics to Offer?	Samuel M. Matsa	1
Computer Graphics Hardware Techniques	D. R. Evans	7
Computer Graphics Software Techniques	S. Bird	17
Interactive Software Techniques	A. R. Rundle	29
Computer Display System Tradeoffs	Harry H. Poole	41
Computer Graphics in the United States	C. Machover	61
The U.K. Scene	F. E. Taylor	85
Low Cost Graphics	Murray A. Ruben	91
Remote Display Terminals	R. Elliot Green	99

PART 2
Applications, Installations

Graphical Computer Aided Programming Systems	C. W. Gear	109
Alphanumeric Terminals for Management Information	A. E. P. Fitz	119
Computer Graphics used for Architectural Design and Costing	P. E. Walter	125
Graphical Output in a Research Establishment	F. M. Larkin	135
Appendix to Graphical Output in a Research Establishment		141
High Energy Physics Applications	P. M. Blackall, J. C. Lassalle, C. Vandoni, A. Yule	149

Mechanical Design Using Graphics	B. T. Torson	161
Electronic Design Using Graphics	D. C. McDouall	169
Man-Machine Co-operation on a Learning Task	Roger A. Chambers, Donald Michie	179
Some Hardware, Software and Applications Problems	R. W. Prowse	187

PART 3
For the Computer Technologist

Present-day Computer Graphics Research	R. Elliot Green	207

PART 4
For Reference

Some Commercially Available Computer Graphics Systems	217
Glossary of Computer Graphics Terms	231
Consolidated Bibliography	239
Subject and Author Index	243

List of Illustrations

	Page
Simple graphics layout	8
Layout with local memories	8
C.R.T. display of electricity sub-station circuit	18
Line categories for windowing	20
Example of expansion	21
Program structure	22
Ring structure representation of rectangle—ambiguous	23
Same representation—unambiguous	24
Display after program request	26
Three-part off-line application program	30
Graphical program structure	30
Menu of symbols on C.R.T.	32
Tracking pattern for lightpen	33
Algorithms for line display	34, 35, 36
Frame scissoring	37
Three typical computer display configurations	43
Two configurations for category selection with computer refresh	45
Three types of word format transmission	48
Multiple display system	55
Display features and applications	56
Three types of optical superimposition of background data	58
Drawing by General Motors DAC System	63
Textile design with graphic console	64
System Devlt Corp. data base display	67
Closed-loop system display	69
Open-loop display (pipeline & tank circuits)	70
IDIIOM—integrated graphic console	76
Graphic console block diagram	77
Early system using storage C.R.T.—the Teleputer	78
Graphic input devices summary	80
Low cost display refresh comparison table	92
Small graphics system layout using general purpose computer	94
Graphics arrangement using satellite computer	94
Layout of remote stand-alone terminal system using dataphone link	95
Programming preparation	110
POLY routine	115
Text communication with ILLIAC II	115
Division of building into functional groups	127
"Blob" shapes for internal building dividers	128
Example of type of plotter output of architectural design	131
Culham Laboratory hardware arrangement	138

	Page
Magnetic field lines—tennis-ball-seam conductor	142
Paths of particles	142, 143, 144
Correct and faulty computation of particle paths	144, 145
Plotting of "phase fluid" formations	146, 147
Stereo photograph of particle trajectories	150
Using lightpen to measure points in bubble chamber event	151
Display of magnified track image	152
Stereo view displays of spark chamber event	153, 154
Display of complex function generated by "function calculator"	155
Display of key actions program to generate function	156
Geometry's role in mechanical engineering	166
Structure of AUTOGRAPH System	170
Display during equaliser filter design	173
Line printer drawing of wiring layout	175
Display of printed circuit board	176
Remote control with animated pole and cart display	182
Manual control with task-independent representation of control problem	183
Automobile silhouette from incremental plotter	210
State diagram for rubber-band line drawing	214
Sofa design from 3-D objects called up	215

PART 1
Systems, Equipment, Techniques, and Trends

What Has Computer Graphics to Offer?

SAMUEL M. MATSA
Manager, IBM New York Scientific Center

The computing capability of second and third generation computers has far outdistanced the growth rate of accompanying input/output equipment. This statement is not intended to minimise the increased efficiency and flexibility of card readers, tapes, discs, key punches and printers. However, it is meant to point out the need for convenient new means of communicating with computers.

Two new and in some ways related developments have given computers a new dimension: time-sharing and graphic data processing. Time-sharing makes it economically feasible to give a single user access to a large computer on an immediate, local basis with quick response time and fast turn-around. Graphics allows the user to communicate with the computer conveniently and in his own terms. This makes it possible to provide to the computer information in its most natural form. Furthermore, it allows the computer to communicate information to the user in a form which is compact, descriptive and most appropriate for a given application. The application areas of computer graphics span the entire spectrum from engineering analysis and design to mathematical analysis and data reduction, and last but not least, to the use of computer graphics as an aid to computer programming and debugging. The other chapters in this book deal with the various aspects of computer graphics in some detail. The purpose of this chapter is to provide a basic frame of reference and some common background for the whole book. It will review some of the history, outline the advantages, describe the major concepts involved and indicate the areas most suitable for the application of computer graphics.

One can trace the beginning of computer graphics to the Jacquard loom where for the first time a digitised representation of a graphical form in punched cards was used to control a loom. Modern computer graphics had its beginning at the M.I.T. Lincoln Laboratory where Dr. Ivan Sutherland developed the SKETCHPAD program which allowed the user to sketch on a cathode ray tube. SKETCHPAD opened the horizons for industrial developments by illustrating the feasibility of having a designer construct rather complex diagrams with the aid of a computer. This work was first published in the Spring of 1963 at the Joint Computer Conference of the American Federation of Information Processing Societies.

What has happened in the five years since then?

A number of different hardware systems have been developed ranging from simple scopes with only text printing capability to elaborate graphic displays equipped with function keys, light pens, and keyboards where one can draw arbitrary shapes in two and three dimensions.

In the software programming area Sutherland's SKETCHPAD has been followed by the development of various programming systems of different levels of flexibility utilising graphics to enhance their effectiveness and/or efficiency. The programming systems provide basic assembly level graphic languages as well as higher level language graphics such as FORTRAN and PL/1. In addition, sometimes application programs include problem-oriented languages and console procedures. Programs have been developed for particular application areas such as drafting, engineering design, statistics, data reduction, optics, etc.

The key to the importance of graphic data processing lies in the additional capability graphics adds to the computer's ability to perform applications. Graphic data processing is a new dimension which has been added to computers. With the capabilities of graphic data processing, a large number of new application areas can be addressed by computers. In addition, a large number of traditional computer applications are enhanced by adding graphics capability.

The advantages and importance of computer graphics for enhancing traditional computer applications is sometimes overlooked. Because historically graphics were looked upon as a technique for allowing the implementation of computer applications which would not otherwise be feasible, they tend to be associated with large systems only. This in turn has discouraged many potential users from applying computer graphics to areas where they could get immediate benefits. Published case studies have shown that when used appropriately computer graphics not only reduces the turn-around time for solving a problem but also cuts down the total cost for a particular problem solution. A specific example was in one of the satellite runs in the U.S. military program. The problem was that of data reduction, analysing data coming back from a satellite run. Without graphics, it took one month to get all the calculations done and to get a result. With the use of graphics, it took only 48 hours, and computer time was cut from about 17 hours to about 8 hours. This was where seeing immediate results allowed one to do a much more efficient programming job.

Graphic data processing provides a common language of graphics and alphanumerics between the man and the computer. A man normally thinks in terms of sketches, drawings, graphs, letters, characters, and numbers. A computer operates in terms of bits, bytes and registers. This makes it difficult for the man to communicate with the computer. In the past, the burden has been on the man; namely, he has had to convert all of his ideas and thoughts to letters, numbers, and a few special characters. The computer, in turn, conversed back with the man in the same medium. With the advent of graphic data processing, the man can work in the medium he understands best; the computer can continue to work in the medium it understands best, with the graphic display console acting as an interpreter between the two. This new dimension in man/machine communication has proved to be of value in applications where:

> Graphic representation is of assistance in the performance of the application, or
> Rapid turn-around time is required, or
> Human imagination, judgement, or experience is required in the solution of the problem.

It is important to realise that computer graphics can be applied to many areas outside what is normally accepted as scientific application areas. There are many potential uses of computer graphics in business data processing as well. For example, a financial analyst can use a graphic display to see the market trends for a particular stock, provide his judgement on the expected value for a period of time, and observe on a real time bases the deviation of the actual fluctuations of the market from the expected range of highs and lows. Then, in turn, he can modify the expected range of values appropriately and thus provide a more accurate and realistic projection.

As another example, people working in urban planning are starting to experiment with the use of terminals where they can get a map of a particular area displayed, and then they can ask for a display of various demographic information such as the spread of a population (e.g. by income level). This can help them to make decisions by allowing a quick evaluation of the current situation.

While computer graphics adds a new dimension to programming it is not in any real way different from conventional computer programming. The various problems and their solutions are similar and analogous to those for many other areas. The availability of graphic higher level languages has made it possible for engineers and scientists to define their problems with computer graphics as easily and inexpensively as they can using a language such as FORTRAN, PL/1 or COBOL. A factor which complicates the problems associated with computer graphics is that practical application of this new tool implies the use of an on-line often real time environment. To practically implement a graphics application, it is necessary to face and solve all of the problems which are inherent in such an environment (time-sharing, multi programming, multi processing, job control, remote job entry, etc.)

A very important area which is currently receiving increasingly growing attention is the problem of modelling and structuring the data to be displayed. This is most important in order to allow the user to efficiently retrieve the information as required. Traditional list processing is only a partial answer to this problem. Graphical data requires much more elaborate structured lists. A number of approaches have been developed, such as plex structures, ring structures, parallel structures. One of the papers presented at the 1968 Congress of the International Federation of Information Processing in Edinburgh, Scotland, described a data structure language and an associated data structure programming system which allows the user to utilise this capability. This work is being carried out at Brown University, Providence, Rhode Island, under the direction of Professor Andries van Dam.

The amount of computer storage required for a graphic representation of a display is quite large even for rather simple displays. Research in this area is currently under way to develop means for storing the representation of graphical displays in large scale secondary storage media such as disks, so that the user can retrieve this information within the time limits necessary to provide for a realistically rapid response for practical use. One of the relatively new techniques which is being explored is known as hash coding which provides for the retrieval of graphical data on the basis of appropriate keys or masks. This work coupled with the concepts of time-sharing may indeed provide at least part of the answer to this major problem area. The structuring of data and its organisation in store are the programming problems that have not necessarily been fully appreciated when people are talking about the potential of computer graphics and why it has not yet been fully realised.

Basically, there are three facets to graphic data processing: the input process, the output process, and the manipulation process. Graphic data processing provides for the input, manipulation, and the output of graphic data, where we define "graphic data" as lines, curves, letters, and characters; basically, anything that can be put on a piece of paper using a pen, a pencil, or a typewriter.

There are a number of devices available to assist in the input, manipulation, and output of graphic data. These include: display consoles, tablets and scanners for input; consoles for graphic manipulation; printers, plotters, drafting machines, recorders and display consoles for output.

Another way of classifying computer graphic displays is across the three facets of graphic data processing in terms of their functional capability. Thus, one can distinguish:

1. Passive output displays *v.* interactive input/output consoles. In the output category one would include printers (which have been used for the generation of schematic diagrams) plotters, as well as, sophisticated drafting machines. The interactive consoles are generally implemented by cathode ray tube displays.
2. Alphanumeric *v.* full graphic displays. Both types include a range of capabilities. Alphanumeric devices often provide for upper and lower case characters. Experimental devices provide for characters of different fonts and multiple size and intensities. Graphic devices allow for the drawing of points and vectors in any direction. Some units are equipped with hardware capability for rotating the display in three dimensional space.

3. Small screen individual *v.* large screen group displays. Most of the units associated with computer graphics are of the small screen variety. Large screens have been used primarily for military applications.

Computer manufacturers have developed experimental systems of graphic equipment families which include a scanner for input, a film recorder for output, and a C.R.T. console for man/machine interaction. None of these systems has so far survived the test of time and/or grown to full-blown commercial use. The primary reason for the lack of success of graphic equipment families has been the enormous programming task of providing software support for such equipment. The outstanding case in this context is the programming required to scan and recognise pictorial information. For example, the lack of techniques for "reading" an engineering drawing by computer, constructing a data structure with sufficient detail to reproduce this drawing and making the necessary associations of graphic and dimensional information. Quite a bit of work is being done in this area and undoubtedly the future will bring additional achievements.

It is expected that other future developments will include the production of colour displays as well as improved means for producing three-dimensional displays. Current research with computer generated holograms as well as work with stereoscopic projections seems to show considerable promise.

The wider utilisation of computer graphics in areas where immediate economic savings can be realised will give impetus to an expanded growth which will bring about the development of inexpensive graphic terminals as well as special purpose terminals designed for particular application areas such as text editing and composition.

There is general consensus that graphic data processing has brought new capabilities and new dimensions to computers. In particular, it greatly enhances the facility for man/machine communication and interaction. The ability to provide instantaneous man/machine communication makes the solving of problems with graphic data processing a much more rapid process with significantly reduced turn-around times. The capability of using graphics as the interface between the man and the computer allows the two to converse in a language most natural to each.

Finally, it lets the man and the computer both operate in an optimum manner. The computer gives us the capability of high speed, raw power, and low cost per calculation. The man contributes problem solving experience, engineering judgement, and imagination. In any application, there is a certain amount of man required and a certain amount of computer. Through graphic data processing the best of these two can be combined in the solution of a wide range of computer applications.

APPLICATION AREAS FOR GRAPHIC DATA PROCESSING

Schematic and Dimensioned Drawings.
Verification Drawings of Numerical Control Tapes.
Electrical, Mechanical, Structural, and Civil Engineering Drawings.
Weather Maps.
Contour Maps.
Layout Drawings for Printed Circuits.
Layout Drawings for Petroleum and Chemical Processes.
Unit Operations Drawings for Petroleum and Chemical Processes.
Pole Line and Distribution Drawings for Electric Utilities.
Exploration Maps for Petroleum and Mining.
Subdivision and Construction Layouts.
Computer Aided Design Systems.
Ship, Aircraft, Missile and Satellite Course Plotting.
Cartography and Hydrographic Plotting.
Pert Network Drawings.
Flight Test and Engine Performance Graphs.
Business Graphs.
Telemetry Data Plotting.
Highway Cut and Fill.
Research and Engineering Data Reduction.
Temperature and Pressure Distribution Drawings.
Fourier Analysis.
Antenna Scatter Displays.
Optical Ray Tracing.
Calibration Curves.
Strip and Disk Chart Analysis.
Mathematical Studies-Function Analysis.
Multi-Dimension Analysis.
Kinematic Analysis.
Route Layout Simulation.
Reservoir Sizing.
Power Spectral Displays.
Quality Control Displays.
Oceanographic Charts.
Oil Production Maps.
Cockpit Visibility Studies.
Aircraft Landing.
Visibility Studies.
Wave Research Drawings.

Biographical Note

S. M. Matsa received the B.S.EE. degree from Purdue University in 1955 and the M.S.EE. degree from M.I.T. in 1956. He joined I.B.M. in 1957 as Project Manager responsible for the development of the AUTO-PROMT system. From 1963–65 he was Manager of Advanced Engineering Applications and since 1966 has been Manager of the New York Scientific Centre.

Computer Graphics Hardware Techniques*

D. R. EVANS
Royal Radar Establishment

Introduction

The purpose of this chapter is to outline as simply as possible, and from a logical rather than a detailed circuit point of view, some of the hardware aspects of computer graphic terminals. This will include some of the more commonly used techniques for the generation and display of alphanumeric information on Cathode Ray Tube (C.R.T.) displays, the principles of operation of some input devices and an appraisal of some display mechanisms in relation to their suitability for use as acceptable computer graphic terminals.

Design Principles

Information to be displayed is stored in digital form and data relating to the format and information content must be passed to the viewing unit back up equipment. In the back up equipment the digital data is converted into analogue waveforms which are used to drive the C.R.T.

The C.R.T. directs a beam of electrons from a "gun" at one end to the screen at the front. A special coating inside called the phosphor, glows under this bombardment, and a spot of light is seen, whose colour depends on the phosphor. The input signals are then used to deflect this spot horizontally and vertically, and a "bright-up" signal is also fed in to switch the beam on only where a visible display is required.

As the beam moves, the spot of light moves over the screen, and in the usual C.R.T. it is necessary to repeat the pattern over and over again at high speed if a steady and coherent picture, and not just a moving spot, is to be seen. This refresh rate is usually 30 to 50 times per second depending on the phosphor. Any reduction in specified refresh rate leads to undesirable flicker.

The operator will usually require facilities for entering data into the computer and this entry process should contain provisions for correction or editing the input message.

This simple concept is shown in Figure 1.

In large systems there will be multiple operator positions and the system shown in Figure 1 can be extended by either distributing the analogue waveforms to the display positions or the distribution can be in digital form with each position having its own back up equipment. In this simple concept the information to be displayed on all displays is usually fed to highways connected to each display and each display

* © H.M.S.O.

has its own bright up signal which is switched on when the data being generated at a particular time is to be shown on that display. Since the refresh rate is determined by the main processor this concept produces constraints in per position capability but economic considerations may dictate this solution.

A more powerful solution in large systems is to provide at each operator position a local memory and viewing unit back up equipment. This is shown in Figure 2.

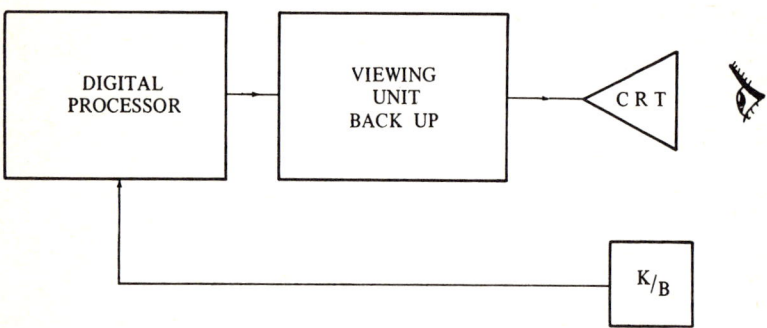

Fig. 1 Simple graphics layout

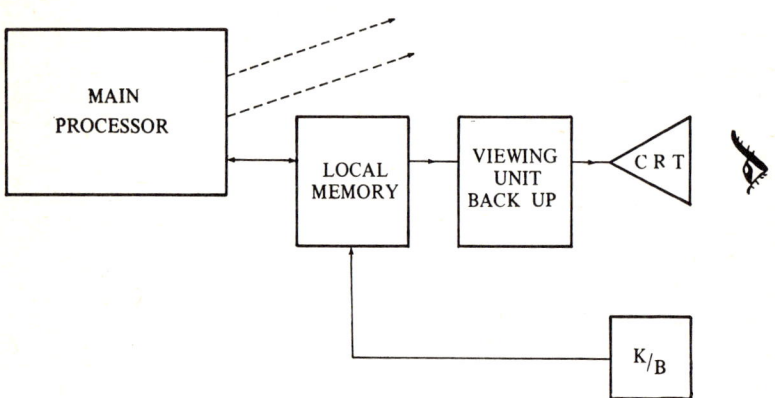

Fig. 2 (K/B denotes keyboard and function keys)

This concept reduces the work load on the main processor because data transfer to the local memory is only required when update is required. The task of cycling the data for display purposes is now a task for the local memory. This local memory can be a delay line, disc, drum or core store. It might even be a passive storage surface.

The advantages of this approach are that it allows the optimisation of display parameters and provides very flexible facilities for amendment and editing of input data. Usually the per position facilities are enhanced in this approach to the point where the local memory becomes a small computer in its own right. The optimisation of the display parameters usually lead to requirements for fast character generation and fast deflection circuitry with consequent wide bandwidth properties. Short wire lengths are desirable and this requirement has a chance of being met in this approach because the circuits can be made as part of the viewing unit. The cost of such a system is seldom small.

One last point worth stressing is that in meeting overall display requirements it is advisable always to examine results in relation to the user requirement and not the engineers evaluation of the user requirement. There is a tendency to seek perfection and this can be very expensive. The user will often not see effects that are very obvious to a display engineer.

Character and Vector Generation Techniques

These can be considered under three basic techniques:

1. Shaped beam
2. Scanning systems
3. Stroke writing systems.

Shaped Beam techniques are represented by the Charactron and Matricon. In each case the electron beam is passed through a controlling template that effectively produces an extruded beam shaped in the form of the character required.

In the Charactron the template contains all possible characters and character selection is carried out by deflecting a beam of electrons to the appropriate position on the template. The electrons that penetrate the gaps are then brought back to the centre of the main deflection system and deflected to the position on the C.R.T. faceplate in which the character is required.

The Matricon uses a template of (say) 7 × 5 holes each of which has a controlling electrode. The computer activates the appropriate holes and a dot picture is produced—after deflection by the main deflection system to the position required on the C.R.T. faceplate.

The Matricon can be considered to be marginally more flexible in that the character is defined by the activation pattern called for by the computer and can thus be changed by program change.

Both these tubes can be considered as fast character generators but care must be taken in not over driving the phosphor, which could affect legibility and life of the tube.

Scanning techniques can be further broken down into Monoscope and Picture scan. In the Monoscope again the characters are defined on a template, usually by a carbon resistance supported on a metal backing plate. Selection of a character requires the centering of a small T.V. scan to encompass the required character and the difference between the secondary emission characteristic of the metal and carbon will produce a T.V.-like video signal. This signal is used to modulate a similar small T.V. scan on the main display C.R.T. This scan is placed in the correct position on the display C.R.T. face by signals passed into the display C.R.T. deflection circuits.

Picture Scan techniques encompass the use of T.V. scan techniques to write a line of characters for tabular display purposes on a line by line basis. The processor has to define bright up signals in a time sequence to write the line elements required. This repeated for (say) seven lines defines a row of characters.

Both these techniques give a low beam utilisation i.e. the phosphor activation time is small compared with the scanning time. The main virtue of these techniques is that of low cost.

Stroke writing systems can be subdivided into Lissajous figure, dot matrix and dot vector systems.

In the Lissajous figure systems, combinations of oscillations of varying frequencies in x and y directions together with bright up signals can be made to form symbols. A simple example is that a waveform of frequency f in the x direction combined with one of frequency $2f$ in the y direction and constant bright up would form a figure eight. Variable writing speed can be a problem and this system is not widely used.

Dot matrix is probably the simplest form of display. Usually a matrix of 7×5 positions is used and a combination of points within this matrix format is painted in sequence to produce a visible character.

The dot vector method is an extension of this method. A combination of points is defined for each character and the electron beam is made to traverse a path from point to point, thus drawing out the character shape. This is the most popular character generation system and is offered by most computer graphic manufacturers. The design problems occur in generating deflection waveforms that start and finish properly to avoid distortions of the character shape. This problem becomes more difficult as the character generation time is reduced.

Input Techniques
Keyboards

Keyboards associated with a graphical display position usually provide a full alphanumeric keyboard and some form of function editing keys. Such a system converts the depression of the appropriate key into a unique digital signal for input to the computer. Verification of the input message necessitates display of the message on the C.R.T. face, and control to ensure that the message is only executed, after verification, by the use of some function key. This system is flexible but slow in comparison with other methods.

Rolling Ball (or the Joystick, which is similar in principle) usually controls digital counters in x and y axes which are decoded to display an electronic marker on the tube face. By rotating the ball in its mounting to position the marker, and operating a function key, positions can be stored or lines or characters can be drawn at the selected position.

(Usually the start point and end point are defined and a vector function key operated to define a line.) This is a well used method and is offered by most manufacturers.

Light Pen is a more natural action. The light pen is a photo sensitive detector housed in a suitable case. This can be used in a number of ways. It can be used to mark positions by merely placing the pen over the area of interest and operating an appropriate function key. When the computer refreshes the area at which the light pen is pointing, the pen senses the flash as the phosphor glow is renewed. This signal is then used to inform the computer that the operator was pointing at a particular part of the pattern drawn on the screen, and the program can then take appropriate action.

To draw with a light pen it is usual to collect from a corner of the display a mark and "tow" it across the display surface. Operation of appropriate function keys allows the definition of the start and finish of vectors or the placing of symbols in specified locations.

The Rand Tablet is very similar in convenience of use. A surface containing a matrix of conductors in x and y coordinates (x and y insulated from one another) is used and this surface is activated by a stylus which generates separable signals in the x and y planes. These signals are input to the processor by the use of function keys and the input is usually displayed on the cathode ray tube for verification.

There are many experimental versions of this concept, e.g. Conducting glass overlays have been used in much the same way as the Rand Tablet. There is a great future in such pointing devices and the nearer the operator action approaches the natural act of writing or drawing the more successful is the man/machine relationship.

Touch Wires

An even more natural action is demonstrated in the use of Touch Wires. This entry system has been developed for use with Tabular displays. The principle is to embed in the tube surface or overlay wires which form part of a bridge circuit. The computer plays out data positioned in such a way that each item is displayed immediately above a touch wire.

The mere act of touching a wire with a finger unbalances a bridge circuit and the computer is thus informed of the item designated. The same wires can then be re-labelled, by the computer, and used for further selection or amendment.

This system is limited in that it can only be applied to operations in which the successive actions follow a logical, and predetermined, sequence. If this is accepted it is very fast, easy to learn and very easy to program.

Display Techniques

Most display systems use a digital store driving a conventional C.R.T. Such systems pose problems in relating display content to acceptability of presentation and lead rapidly to compromises on refresh rate, character generation and deflection times, brightness and practically every other display parameter.

The need to compromise has led to many innovations, e.g. Main and Secondary deflection systems, the Main systems being comparatively slow but working over the whole C.R.T. face, the Secondary system being fast but only operating over a limited area. This system, though well proven, is already proving not flexible enough and Main deflection systems are now expected to be fast. This costs bandwidth and creates many circuit problems. The Matricon and Charactron, combined electrostatic and electromagnetic deflection, rear ported optical projection of static data and many other techniques are the tools used by the display engineer to meet the need to display more data on a given display whilst still maintaining an acceptable presentation.

The combination of digital store and conventional C.R.T. is flexible and very effective and there can be little doubt that for the immediate future this is the only readily available solution. Improvement in the C.R.T. is still needed. Phosphor research is a topic that could produce major impacts. More efficient phosphors with a "drop dead suddenly" persistence characteristic to ease refresh rate problems, resistance to burn to increase tube life and colour as a coding mechanism are desirable attributes for the next generation of display C.R.T.'s. Cheaper digital stores, and cheaper small stores at that, are desirable to bring down the cost.

Alternative Approaches

An anomaly exists in the above solution to the display problem in that the processor carries a high repetitive work load entirely for display purposes. This work load bears little or no relation to the operator response time. For this reason attention must be given to other local storage mechanisms.

Scan Conversion and T.V. Monitor

This is readily available at this time. The principle is to play out from the digital processor only when update is required.

Character generation and function generation can be comparatively slow and the output for display is written on to a storage surface by well tried methods. The information is scanned off the storage surface in T.V. fashion and displayed on a T.V. monitor. (The storage tube is a display "slow-in fast-out data rate converter".) The decay characteristic can be controlled by controlling the erasure characteristics of the Scan Converter. Thus mixed decay (or persistence) characteristics can be achieved by using multiple scan converters driving a single monitor tube.

Resolution is poor and improvement must be sought in improved electron guns. (And in improved electron guns working at low accelerating potentials.) If this is achieved scan standards must be changed and this poses interesting circuit problems. Increased bandwidth is required both in the scan converter equipment and the monitor display and this is costly. At the moment this approach is very expensive for what it provides and is used mainly in radar data processing systems where there is a need to combine low data rate inputs (radar) with computer generated data (track labels).

Direct View Storage Tubes (D.V.S.T.)

The D.V.S.T. is akin to scan conversion in concept except that in the D.V.S.T. the information for display is written as a charge pattern on a storage mesh immediately behind the viewing surface. A separate electron gun floods this mesh with low velocity electrons and where a charged area is written the electrons penetrate and are accelerated to the phosphor and appear as a visible display.

The display is very acceptable. It is steady and bright and reduces the work load on the main processor. Resolution is marginally good enough and good page erasure can be achieved. Selective erasure, i.e. the ability to change one item without altering other displayed data, is desirable. Experimentally this has been achieved and it is possible to write, erase and to some extent show non stored marks on a single display device. The development of this so called "multimode working" D.V.S.T. is essential, though good page erasure displays are available now.

Dark Trace Displays

In many ways these displays can be regarded as meshless D.V.S.T.'s. The phosphor does not emit light when bombarded by the electron beam—it discolours and can then be viewed as a high contrast display when irradiated by a light source. Resolution is good. Writing speed and contrast not quite good enough but erasure (and only as page erasure) as yet is very slow. The potential advantage of these displays lies in the possibility that, if the erasure problems can be solved, they could provide a low cost D.V.S.T.

Matrix Displays

All the previously mentioned display devices use an electron beam in some form of glassware. C.R.T's are bulky and require, in general, analogue drive waveforms. The appropriate back up equipment is complex and can be quite expensive.

Matrix displays are potentially a better match to the computer output. Theoretically a flat display surface, addressed directly in x and y planes, would be a delightful tool for the system engineer. Much research has been carried out in this field particularly in the area of electroluminescent devices. It should be stressed that some latching device is essential per display element to hold information set in, otherwise all the problems of refresh rate, brightness etc., remain and the ubiquitous C.R.T. is well suited to these problems. The achievement of this latching, the addressing problem to stimulate only the desired display elements and the low efficiency of most existing devices has militated against anything but extremely experimental displays being made.

One mechanism does show promise and that is the use of a gas discharge (neon glow) matrix. This inherently has storage (latching) because the "strike", "maintain" and "quench" potentials of a single element can be made well separated. If a sandwich structure can be made which has cathodes in the x direction, anodes in the y direction and gas contained between these planes in a honeycomb structure it should be possible to make a suitable display panel. The maintain potential is applied between all the cathodes and anodes. If a given cathode is pulsed down and a given anode pulsed up by half the potential difference between maintain and strike potential—only the hole at the intersection of that anode and cathode should strike and emit light. Removal of the pulses would leave the element ionised and emitting light. Quenching would require the same mechanism but with the cathode pulsed up and the anode pulsed down.

The problem is to make such a device. Each element needs a unique series resistor and the technology to achieve this is proving difficult. Resolution and life might well prove to be equally difficult problems.

Conclusion

Computer graphic hardware techniques are well developed for most display systems using conventional C.R.T's associated with digital storage devices.

In large systems problems arising from the need to produce acceptable displays of large data content are already producing significant circuit problems.

Alternative display mechanisms are receiving attention and might well produce significant impact on overall system problems and perhaps equally significant economic savings.

Acknowledgements

This chapter is a condensation of many discussions with and much experimentation by colleagues in the Royal Radar Establishment and in many industrial organisations associated with our work.

It has been contributed by permission of the Director of the Royal Radar Establishment. The copyright is retained by the Controller of Her Majesty's Stationery Office.

Biographical Note
D. R. Evans, B.Sc., graduated from Birmingham University in 1941 and joined the Telecommunications Research Establishment at Swanage, Dorset, in that year. Has remained with the establishment through changes of Ministry, location and name to present day. Has worked in all applications departments and has been associated with data handling problems since 1948.

Computer Graphics Software Techniques

S. BIRD

The Marconi Company

Display Commands

The first step in programming display equipment, in order to produce a picture on the screen, is to generate a stream of binary words. These form a series of commands to the display; each command causing the display to carry out one of its functions, such as: to position the beam; to move it linearly painting a line; or to produce a character. Depending on the make of hardware and the word length, these various types of operation are distinguished by, say, the 4 most significant bits (or by control words which change the subsequent mode of operation) the remainder of the word being the binary values of X and Y, changes in X and Y (each usually to 10 bits), or the codes of 2 or 3 characters.

Display File

Thus a sequence of these words corresponds to a picture. They form what is known as the "display file". This either occupies part of the computer's immediate access memory or a buffer store included as part of the display equipment. Either way, to produce a steady picture, the file must be continually cycled through and passed, one word at a time, to the display controller proper—in the case when the file is within the computer, this is done by Autonomous Access or some form of Autonomous Data Transfer.

Figure 1 shows an example of a picture appearing on a cathode-ray-tube screen. It is the circuit of a Primary Sub-station in the Midland Electricity Board's Birmingham area distribution scheme. Such a diagram is displayed by continuously running through some 800 words of storage. Each element of the picture is refreshed 25 times a second.

Animation

Once a static picture has been produced in this way, it can be "animated" by updating the contents of a few locations in the display file. This might be to alter the length of a vector, the position of a picture element, or to change some characters. The viewer sees the effect of the change to the stored data on the next frame, i.e. near enough immediately. Referring to the M.E.B. diagram again, a symbol beside a switch is added or removed to indicate that the switch has been opened or closed. Numbers beside a conductor alter as the current flow in that conductor alters. In this respect, a display differs from other computer peripheral devices—even a digital graph plotter—and it is this that makes it such a powerful means of man-machine com-

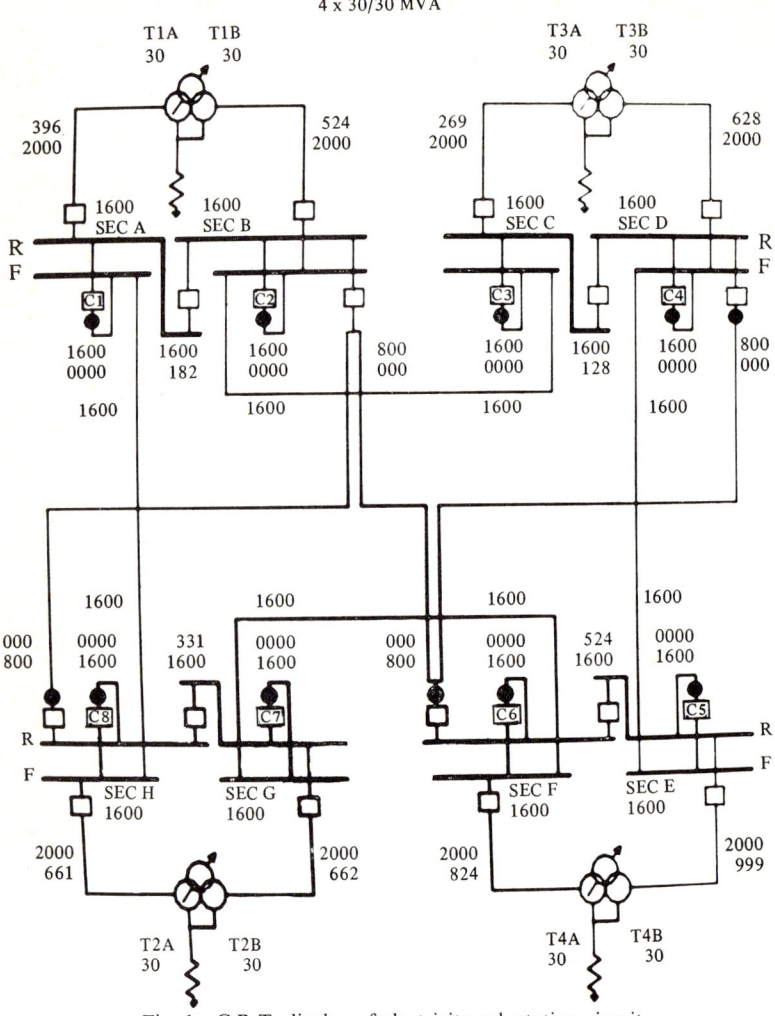

Fig. 1   C.R.T. display of electricity sub-station circuit

munication. The fact that it only needs to be sent the differences, effectively increases its already high speed and also, to some extent, helps to highlight the changes. It produces a truly dynamic presentation of a real-time situation.

Graphic Languages

Graphic languages exist which enable users to build up diagrams, for insertion into the computer memory, by writing a verbal description of the required picture. Such languages, if developed for displays, tend to have had their form dictated by the hardware—this may be rightly so since this leads to an efficient use of all the hardware facilities. In some applications general geometric languages such as APT could be taken as the basis. Assembler programs read punched tapes conveying this verbal description and produce the required bit patterns in the computer or display store. Such assembler programs can give a dictionary output indicating what store locations it has allocated to designated items. Thus other programs can change the contents of these locations to animate the diagram. In general there is a tendency for graphic language assemblers to be unsuitable for fully dynamic work.

Data Structure

In any sophisticated Computer Graphics application, the display file is an almost trivial end-product. If any serious computation is to be done on the system or device represented on the screen (even the production of a parts list) some other form of storage or "data structure" is required as well. The display file is not a convenient form of representation except to the display itself.

A data structure forms a model of the system and may contain more information than the display could show at any one time—and may be of greater accuracy. It is possible, from one data structure, to produce many different display files (and therefore different images). Each would be a different visual representation of the same object—for example, they could be different views, or different types of projection, or part views to differing degrees of expansion. There may be other modes of representation that are not truly pictorial, such as parts lists and schematic diagrams. Thus the display file (or the resultant image) is a pictorial representation of, and the data structure is a model of, the device or system that the man is trying to describe to the computer or the computer to the man.

Expansion and "Windowing"

The data structure can contain co-ordinates to a higher degree of accuracy than that to which the display is capable of working. It follows then, that part of the picture can be meaningfully expanded. When this is done, other parts, of course, disappear off the sides of the screen. The screen then represents a window which can be moved around so that any portion can be viewed. The new ends of part lines are computed where one or both of the true ends are off the screen. It is first necessary to determine whether any part of a line should show. The various categories into which a line can fall are demonstrated by Figure 2. In the system developed by the author, the data is stored to 23-bit accuracy, whereas the display is only capable of 10-bit resolution. This means that expansion of up to 8192 times is possible.

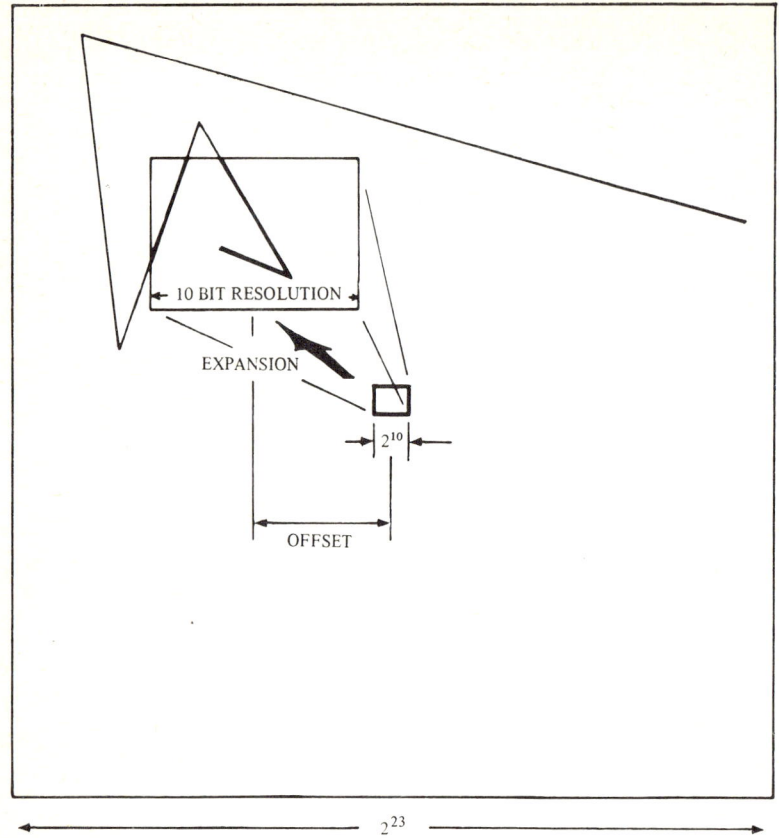

Fig. 2 Line categories for "windowing"

Figure 3 is included to illustrate the effect of having this amount of expansion. Although this may seem a frivolous example, there are important consequences. Since it is equivalent to having a drawing board two miles wide, it is possible, on one drawing, to show for example, all the rivets in a large ship.

Program Structure

The display file is produced from the data structure by the display program package and the transformations such as expansion, rotation, offsetting and "windowing" are carried out in the process. Special Input programs take signals from tracker ball, light pen, etc. (and any keys that indicate the required mode of working) and produce data structure direct. Thus the image on the screen forms a true tell-back to show that the movements of ball or pen have been correctly interpreted (Figure 4).

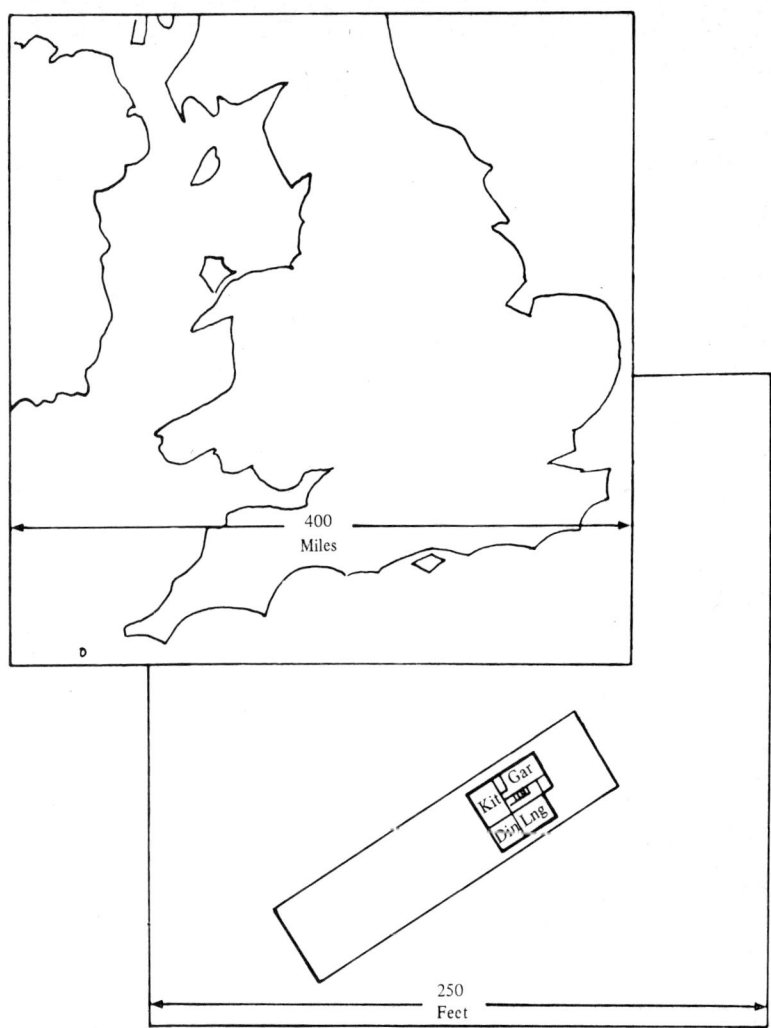

Fig. 3 Example of "expansion"

Additional information can have been deduced during the input process and be evident in the data structure, but not so obvious in the display file. For example, if a mode key were selected so that exactly horizontal or vertical lines are produced automatically, then the fact that the figure is intentionally rectilinear is recorded. Such additional information is also available to any application programs.

Ring Structure

Various forms of data structure exist, but one which is particularly appropriate for display work, and the one adopted by the author is a

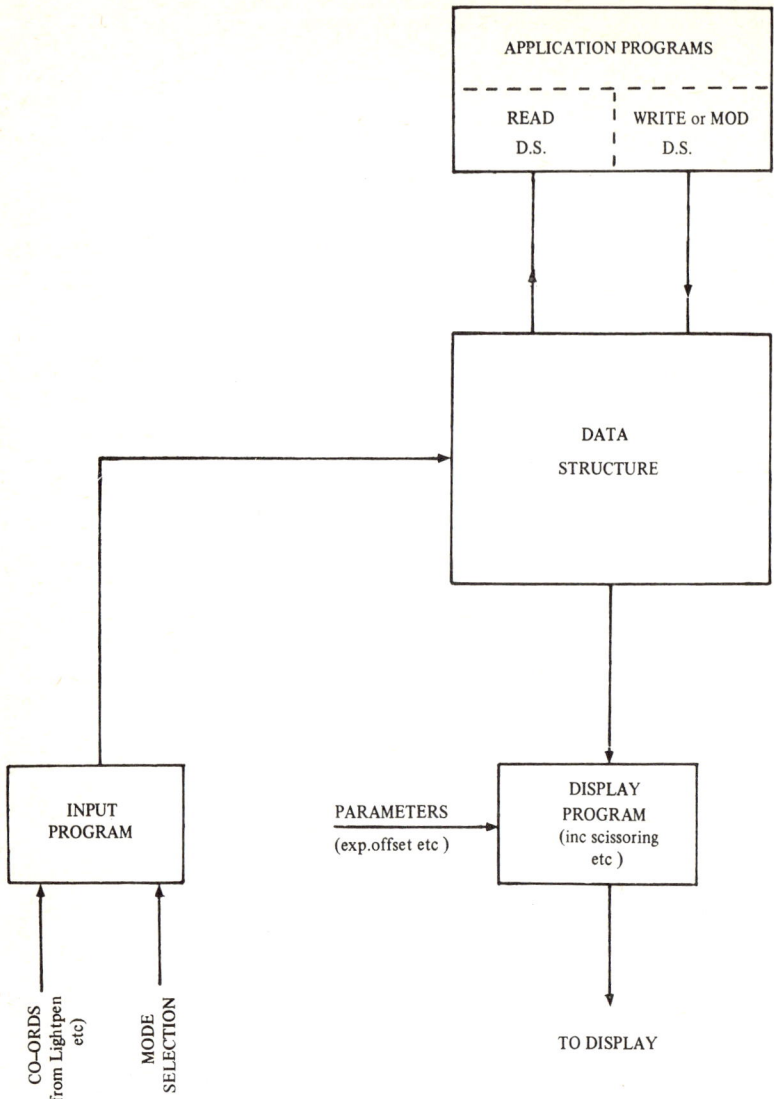

Fig. 4 Program structure

"ring structure". The storage is divided into blocks, not necessarily all the same size, but typically about 6–10 words. Each is assigned to an element such as a line, a point, a drawing, etc. Part of each block, known as the tail, contains specific information about the element such as name and, in some cases, numerical values. The rings are formed by the words in the "head" and these rings associate elements that have a common factor, such as a number of lines terminating at one point.

A RECTANGLE

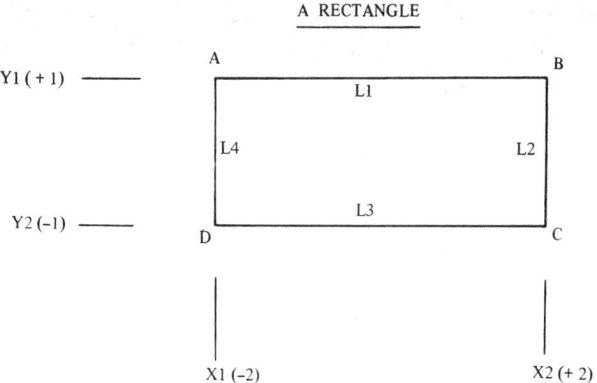

DISPLAY CODING

POSITION	X = -2	Y = +1
VECTOR	ΔX = +4	ΔY = 0
VECTOR	ΔX = 0	ΔY = -2
VECTOR	ΔX = -4	ΔY = 0
VECTOR	ΔX = 0	ΔY = +2

TOPOLOGICAL DESCRIPTION

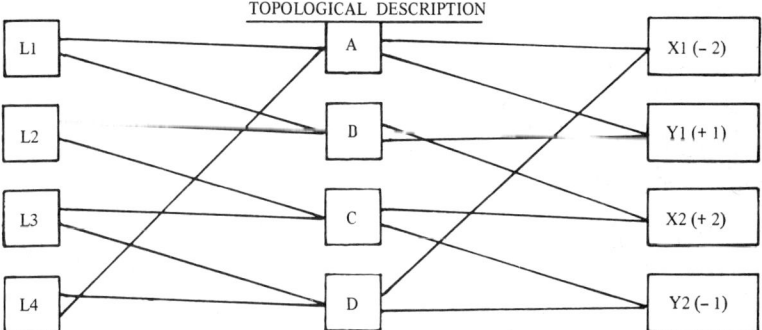

Fig. 5   Ring structure representation of a rectangle—ambiguous

One word in the head of each line-block would give the address of the next line-block, so that they form a ring that can be followed treasure-hunt fashion. The block representing the element that they have in common—the point in this case—is also one link in the chain ring, but is specially marked as the "master" of the ring.

The nature of a ring structure is such that, as seen in the example above, it is particularly suitable for describing the topology of a line drawing (Figures 5 and 6 show a rectangle—and how it would be represented in ring structure). It is not obvious from the format of the display coding in Figure 5 that the four lines form a closed figure.

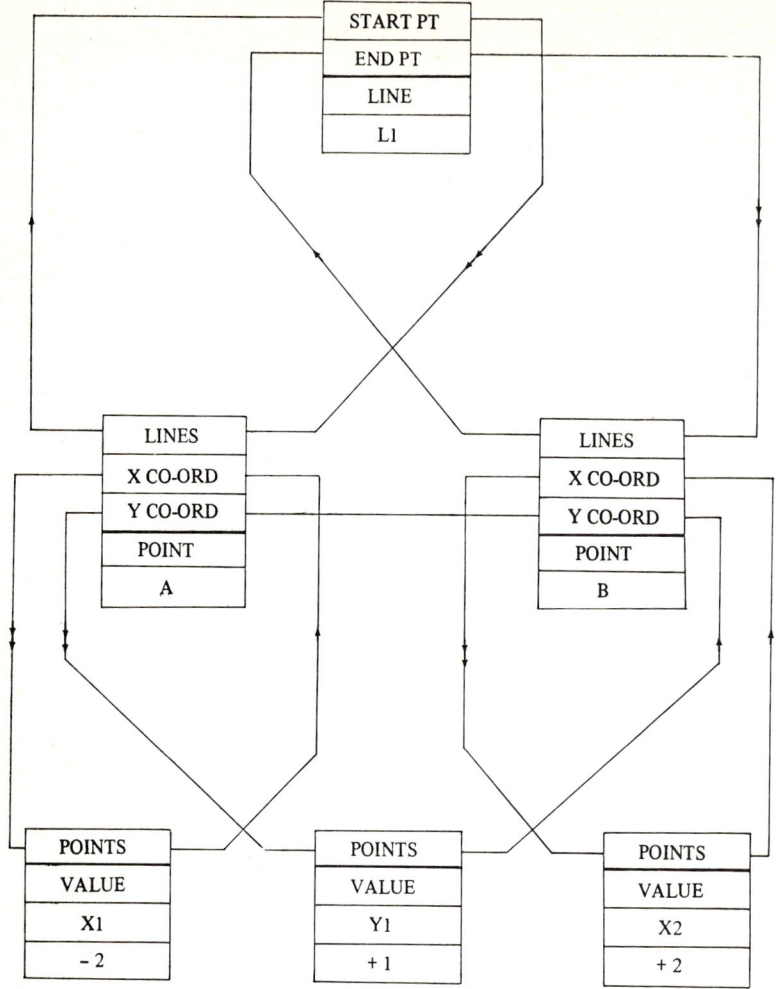

Fig. 6 Ring structure representation of a rectangle—unambiguous

Furthermore, the horizontal lines might have been so by coincidence and it was not the intention of the operator that they should necessarily remain so; or they might have a small vertical interval such that zero is the nearest approximation that the display can give. The ring structure, however, described in Figure 6 is unambiguous.

The topological description shows that $L1$ for instance is between A and B, while $L4$ is between D and A—not a 5th point E which happens to have the same coordinates as A but A itself. Further, since, for instance, B and C share $X2$, $L2$ is inherently vertical. In Figure 6 the single links in the topological description have been replaced by rings.

Picture Manipulation and Deletion

The general principle is to put numerical information in as few places as possible in the ring structure. For example, the Y co-ordinate of a number of points that are on the same horizontal level, would appear in one place only. This has the advantage that it is only necessary to alter this one location to cause the picture to change in a sensible, manner. For instance, by changing the one location, $X2$ to, say, $+3$, $L2$ is moved bodily to the right and the lengths of $L1$ and $L3$ are increased. Thus programs can easily be produced to manipulate parts of the picture, i.e., move elements or alter proportions, whilst maintaining any of its intentional attributes. The ring structure automatically gives consistent answers to questions such as "what points are the ends of this line?" and "what lines terminate at this point?" despite the fact that the number of lines may change from time to time and was not known when the point was first created.

When deleting items, the ring structure makes it possible to find and delete all other items which can now no longer exist. Since the physical positions of blocks in the store is immaterial, the holes left by deletion can be re-used by any subsequent new item that happens to require the same amount of space.

Subpictures

It is possible to quickly produce symbols on a diagram by making "subpictures" of them. This is done by first drawing the symbol as a picture in its own right, thereby creating ring structure that describes it. The data structure of the diagram then has to contain one block for each instance of that symbol and a ring leads from these blocks to the data structure of the symbol original. Each such block contains parameters that are particular to that instance of the symbol such as size and orientation. The display program follows the ring each time—picking up these parameters in passing—and getting the remaining details from the data structure of the original. If the original ever needs to be modified the new version will appear everywhere that it is used.

Application Programs

As suggested in Figure 4, applications programs (i.e. programs written to solve particular application oriented problems) should obtain their data from, and return their results to, the ring structure. This is done via interface programs which extract data and pass it over in the sequence in which it is required. The information thus conveyed may be data (coordinates, lengths, angles, etc.) or "verbal" descriptions (of the topology of the picture or the form of the ring structure). The applications program may have been written in one of the conventional high level languages. Figure 7 shows how the screen would appear immediately after the operator has requested a program (called "BEND") to analyse the bending moments in a beam with a uniformly distributed load. He has named the program and then quoted the names that he has used on the diagram for the parameters of the analysis program. When he drew the diagram, he inserted the output, as well as the input, parameters using dummy values. In this way he has determined the position and method of presentation he prefers for each result. In the example, he has had the array *BM* shown by a graph of vectors, but the value of *BMAX* shown by characters whose position is dependent on *LBMX*. If he now alters the position of the prop and re-runs the program, the position and value of *BMAX* will alter. Thus he could experiment until the peaks of bending moments are roughly equal.

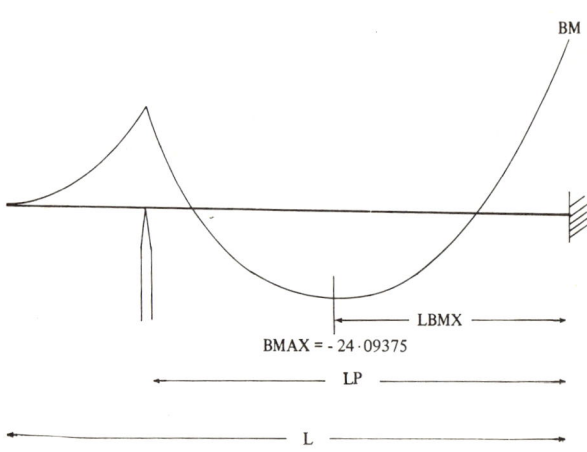

Fig. 7   Display after program request

Operating Systems

A few such programs could be initiated, one at a time, from the graphics program, and control would revert to the graphics program when each one finished. A more sophisticated arrangement that is appropriate when a large library of programs are held in store, puts everything under the control of the Operating System control program. The operator working with the display then has access to the whole library, the programs of which can have been written in a variety of source languages (some high and some low) but are all now compiled into a standard form (the compiler programs themselves will also be in the library). Programs written in any new language, for which a compiler becomes available in the future, will automatically be compatible with the system. The Operating System may also use the same programs to carry out requests received from operators at other positions (not necessarily display positions) or routine low priority tasks.

Biographical Note

S. Bird received the B.Sc.(Eng.) degree in Electrical Engineering from Kings College, London. He joined the Marconi Company in 1947 and has worked first on hardware and now on software aspects of digital equipment. He now heads a team producing the software for the Marconi Graphical/Tabular display.

Interactive Software Techniques

A. R. RUNDLE
Elliott Automation Systems Ltd

Introduction

The Graphical Display provides the most flexible and adaptable means of giving information to and receiving information from, a computer. It has not yet fulfilled its potential because it is probably the most complicated single peripheral to be attached to a computer.

In the last year or two, several manufacturers' software has developed to a stage where the display can be regarded as just one more peripheral to be used with only a little more thought than need be applied to magnetic tape or disc. For this reason, this chapter is intended for those who have not used a graphical display before and who generally use high level languages. The ring structure method described in the last chapter operates at a low level without the user having any detailed knowledge of its operation. The overall system, however, whether using a ring structure or not, will make available to the user a set of sub-programs, or procedures, which he can call upon to generate a picture on the screen, in much the same way as with most digital plotter packages. In addition, there will be a number of routines to enable him to input relevant data from the display, for example the position of the lightpen, the object at which it was pointing, or the button pressed. Some example program statements are shown in Figure 5, which implements the flowchart of Figure 6.

Very few, if any, of the techniques described are new, although to my knowledge they have not been collected together in one place before.

Structure and Layout of a Graphical Program

With most display programs, there are at any time a number of meaningful actions that a user can take, depending on which part of which particular program he is using. When the program has responded to his action, it again waits for him to take further action to direct control of the program to a different routine, and so on.

Instead of the usual three-part structure of off-line "application" programs (Figure 1), graphical programs tend to take a more complex branched structure (e.g. Figure 2) where many routines can be invoked by a central control routine.

This enables the user to draw something on the screen, perform calculations and analyse the results; and if they are unsatisfactory he can modify the picture and perform his calculations anew. This interactive use of the display enables the computer to be fully utilised as a design tool.

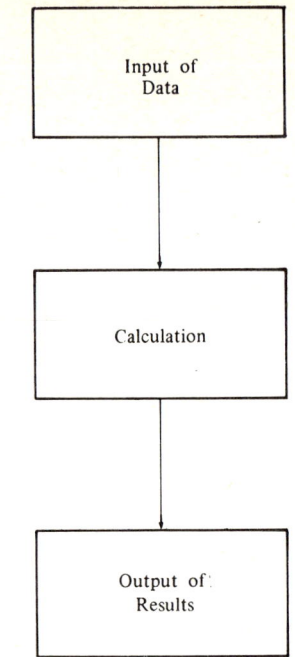

Fig. 1 Three-part off-line "application" program

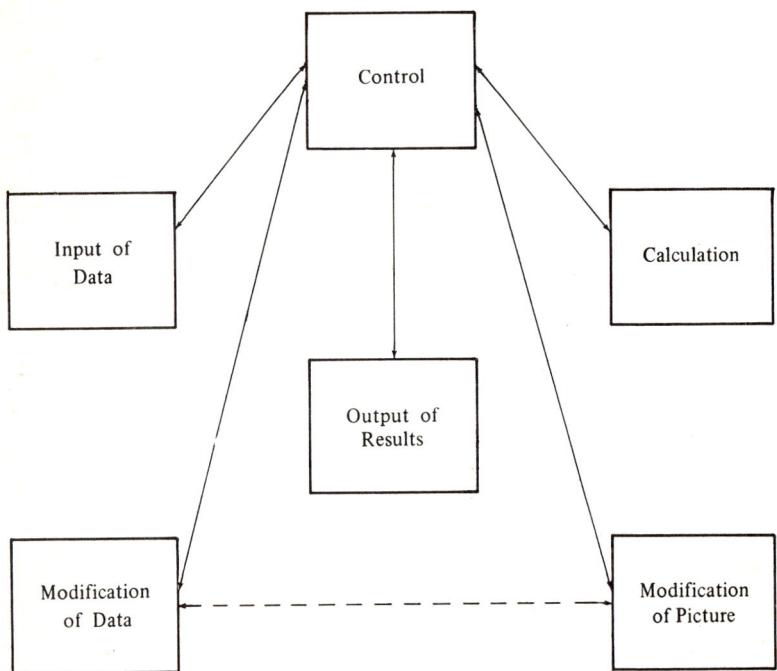

Fig. 2 Graphical program structure

Although the more publicised display programs are highly complex, there are many applications where the display can usefully be used as little more than a fast plotter, and if the display is accepted as a general input/output device for a large number of computer installations it will be because of the comparatively simple applications such as presentation of management or scientific data in graphical form, or for checking tapes for numerically controlled machine tools.

In many cases the graphical interactive part of the program will be trivial compared to the calculation and analysis routine. The majority of graphical programs, therefore, will be written in high-level languages such as ALGOL, FORTRAN or, if it gains universal acceptance, PL1; in fact a large number of existing programs could be converted to utilise graphics in a comparatively short space of time. Some examples of such programs are structural analysis and printed circuit layout packages.

Controlling the Program

The ease of operation of a graphics program, and therefore to some extent its usefulness, depends on the way in which the various input devices associated with the display are used.

The devices normally associated with a graphical display with which the user can control his program are a light pen, a set of keys or a button box, and a typewriter keyboard.

It has been found convenient to arrange for the most common actions such as drawing a line or deleting an object to be instigated by the action of depressing keys or buttons: with only a little practice the operator can use the program without taking his eyes away from the screen. This makes speedy operation possible.

The choice of which function to associate with which button is usually at the user's discretion, and a convenient method of arranging the control of the program is to find out which button has been pressed, and to go to the part of the program appropriate to this button.

It should be recognised that the user is likely to make mistakes, so care must be taken to ignore obviously wrong commands, and to give him a chance to change his mind where mistakes could be disastrous e.g. in a deletion routine.

When in this part of the program, the typewriter can be used rather like a sophisticated button box. However, it must be said that the user is even more likely to make a mistake when confronted with a choice of a large number of keys than when there are a smaller number in the form of a button box.

With complex programs, the operation tends to become too complicated to rely on just a button box, since there are usually not enough buttons to give each one a unique function, and the user has difficulty remembering what to do.

Messages or symbols can be put on the screen to assist the user to specify the more complex operations. By pointing the lightpen at the requisite message, the user can indicate which operations he requires, with far less likelihood of mistakes occurring.

The control symbols or messages to which the user can point are generally referred to collectively as a "menu" (see Figure 3).

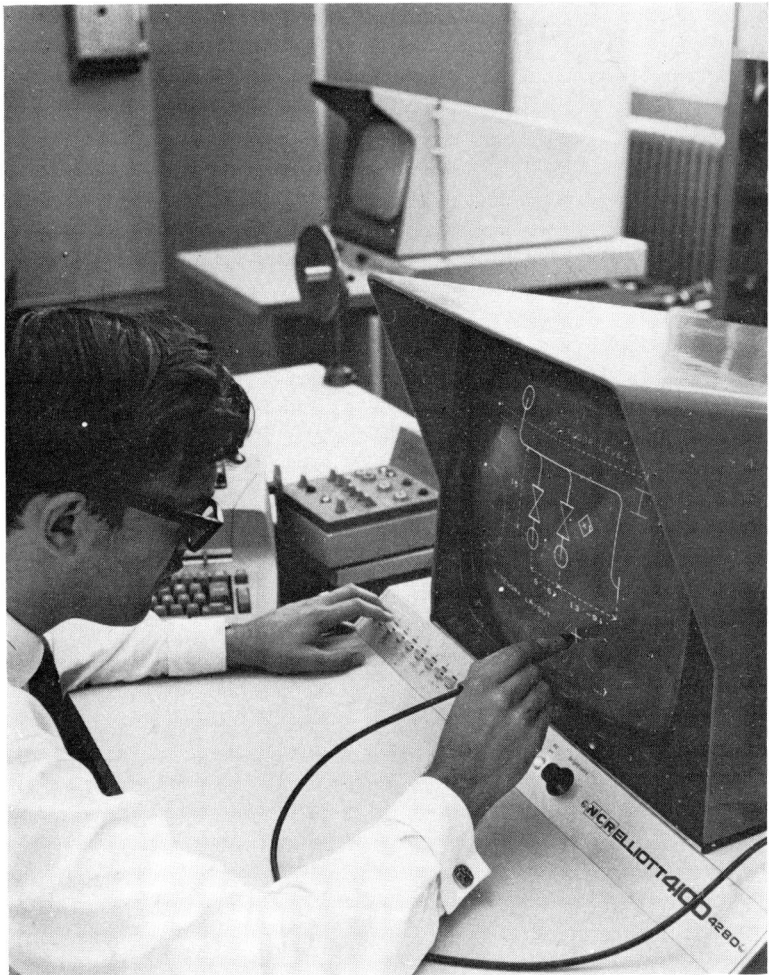

Fig. 3   The "menu" of symbols can be seen at the bottom of the screen displayed with low brightness

The "menu" should be changed by the program whenever applicable, so that the user is not faced with a bewildering variety of choices, as this, too, can lead to a higher incidence of errors.

Pen-Tracking

The field of view of a lightpen is typically at least a quarter of an inch in diameter, and often more when not directly in contact with the screen. The problem arises of using this blunt instrument to define a position on the screen with much greater accuracy, ideally to the resolution of the basic grid with which the C.R.T. displays objects. Some means has to be found of indicating the centre of the field of view, or rather the position that the computer has calculated to be the centre of the field of view; although most manufacturers supply the necessary software to do this a brief explanation of the method involved may be found helpful.

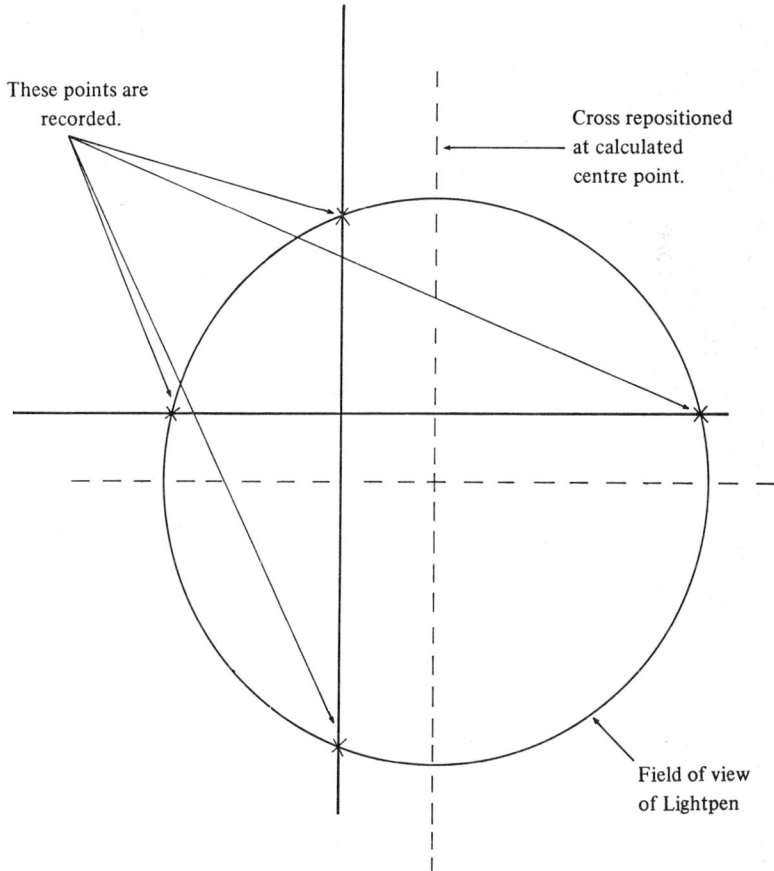

Fig. 4 Tracking pattern for lightpen

A tracking pattern, usually a cross, is displayed somewhere in the field of view of the pen (Figure 4). For each of the four arms of the cross drawn inwards an interrupt is received when the vector enters the pen's field of view, and the co-ordinates noted. From these co-ordinates the centre can be calculated and the cross repositioned. By repeating this

D

process the cross will follow the movements of the pen, continuously indicating the centre of the field of view.

It is possible to lose the "tracking cross" by moving the lightpen quickly; if this happens a spiral search pattern can be displayed around the last known position of the pen. With one or two added refinements it can be made impossible to leave the cross behind.

The lightpen can therefore be used not only as a means of identifying an object on the screen but also for input of co-ordinate information.

Let us examine some of the problems encountered when using the graphical display as an input device.

Lines

First of all, for most applications, the user must be able to draw a line on the screen. There are basically two algorithms for this (Figures 5, 6 and 7). Using the first method, the user would point the lightpen, or otherwise indicate the starting position of the intended line, and then press a button. He would then position the lightpen at the finishing point of the line and press a button again. Having thus indicated the end points, the program can generate the necessary line. The second method continuously re-draws a line from a starting point to the current position of the pen, the effect being of "stretching" a "rubber-band" line until it has the required size and direction.

comment A two-dimensional ALGOL implementation of Fig. 6;
$j = Button$;
comment Waits until button is pressed;
Pentrack (($X1$, $Y1$);
comment Puts current pen position in $X1$, $Y1$;
$j = Button$;
Pentrack ($X2$, $Y2$);
Newbuf (*Buf*);
comment Clear a buffer in which to generate code;
Point ($X1$, $Y1$, 0);
comment Last parameter is zero for invisible, non-zero for visible, point or vector;
Vector ($X2 - X1$, $Y2 - Y1$, 1);
Insert (0, 0);
comment Vector inserted into Display File and hence drawn on the screen;

Fig. 5

"Rubber-band" lines are generally easier to draw initially, but the simple method is possibly better once the user has gained some experience, since lines are often drawn between two existing points. Perhaps the best course of action is to include both methods and let the user choose which one he prefers.

Facilities should obviously be included for deleting and altering lines once drawn.

Gravity fields

Because it is difficult to position the tracking cross both quickly and accurately at the same time, methods have been evolved for putting

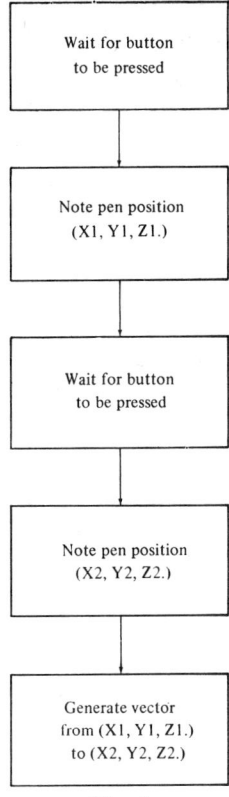

Fig. 6

"gravity fields" around points on the screen, so that the cross will gravitate towards these points when in close proximity to them. In this way, for example, it is possible to draw a line between two existing lines without positioning the track cross nearer than (say) a quarter of an inch from the lines. The method adopted will depend upon the application and the limitations of the software being used.

The simplest form of gravity field is the "grid" technique. Whenever the position of the cross is obtained, it is rounded off to the nearest n units, depending on the resolution required and re-positioned accordingly. This gives the effect of drawing on graph paper, and can be very useful for the type of application where pre-defined items are to be assembled to form a whole, such as with circuit design and analysis, in some architectural problems, and generally where the pictorial representation of data is stylised rather than accurately to scale.

Another straight-forward method is to test if the cross is near one of the end points of the existing lines. If there are a large number of lines this can take some time, but there are cases where this method has to be used. A means of reducing the calculation involved is to test whether

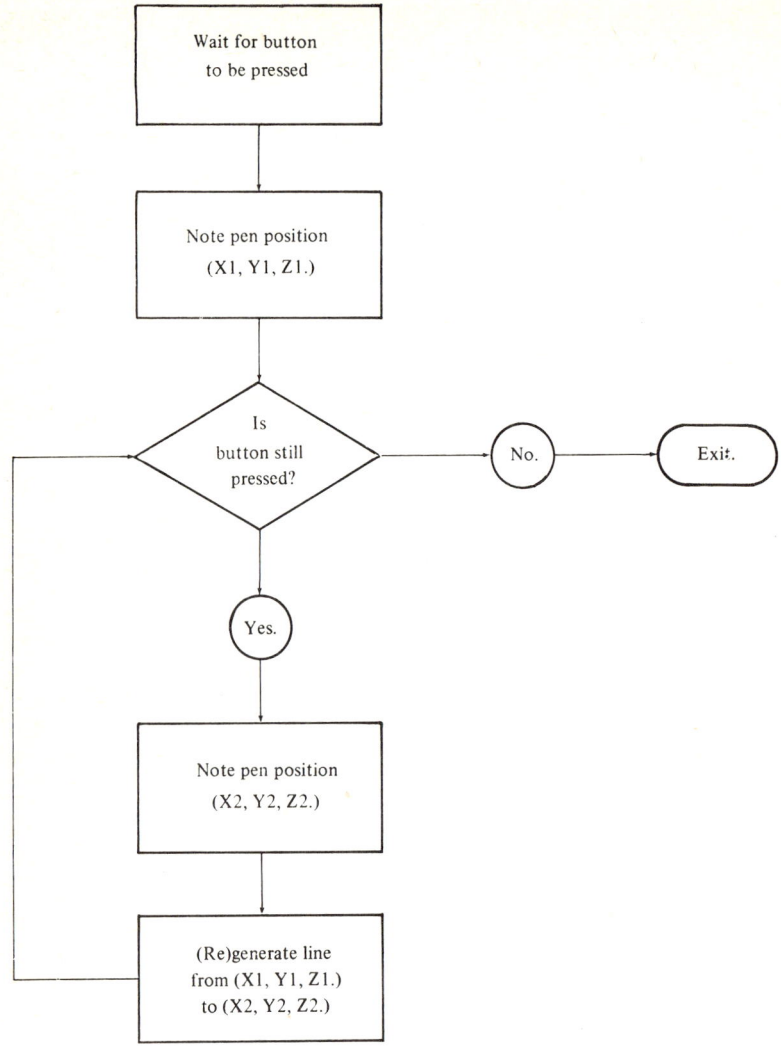

Fig. 7

the lightpen can "see" a line, and, if it can, the gravity field routine need only be performed on the end-points of that particular line.

Gravity fields can be applied to other than lines, for example arcs, ellipses, or even complex subpictures representing transistors or pipe valves. Also, by straightforward co-ordinate geometry, gravity fields can be made much more complicated, in order to put the cross not only on end-points but also along the lines, on intersections, and so on.

It is helpful to the user if the gravity field routine can be performed and the cross repositioned periodically, for example at the end of every frame, but if this is not possible it is usually sufficient to apply the routine only when the position of the cross is meaningful, perhaps when drawing a line or indicating a specific point on the screen.

Frame Scissoring

A common problem encountered is that of representing a picture that would normally occupy the whole of a large engineering drawing-board. The display can be considered as a "window" through which to view the picture, and by moving the window in different directions the whole of the picture can be examined.

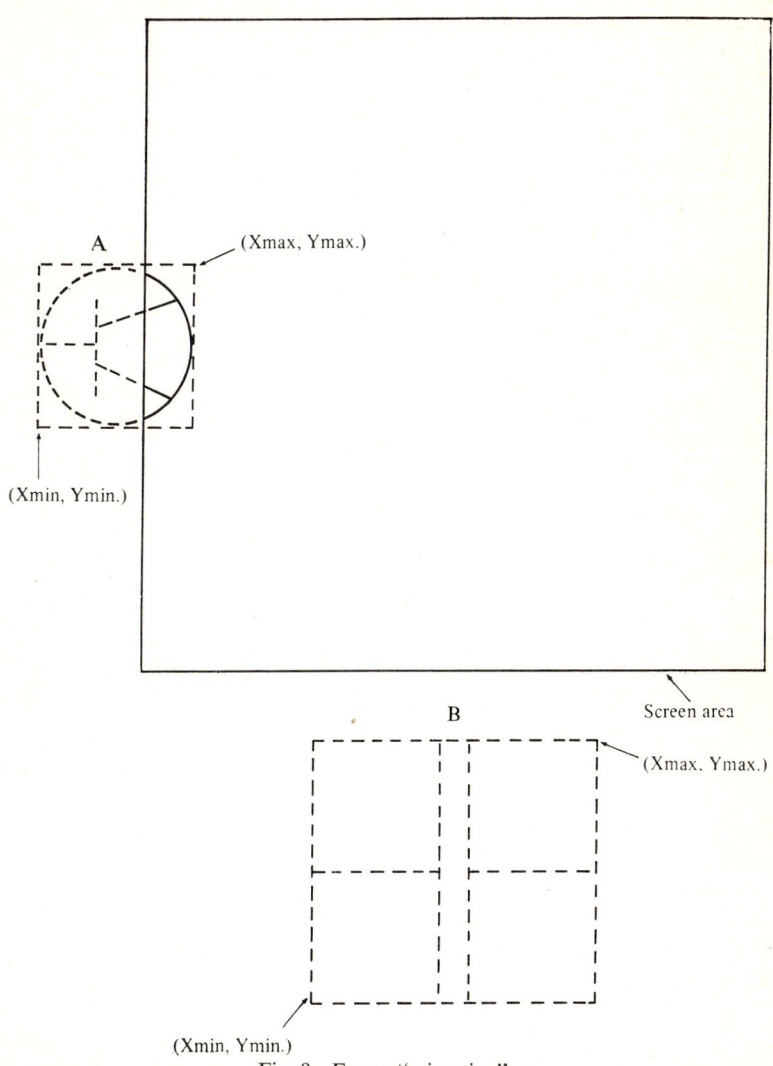

Fig. 8 Frame "scissoring"

The picture outside the screen boundaries cannot be drawn, so on displays without hardware "frame scissoring" the intercept of each line, or other graphical element, with the boundaries of the screen has to be calculated and the viewable picture drawn accordingly.

Theoretically, hardware frame scissoring should solve this problem. In practice, however, the line drawing capacity of displays with this facility is sometimes not sufficient to draw the whole of a large picture, even allowing for the fact that vectors "scissored" outside the screen area can be drawn faster than those actually on the screen. A compromise solution has to be sought.

If the maximum and minimum x and y coordinates of each object drawn are stored the only test that need be made is whether or not any part of the object is on the screen. If it is on the screen, the object is drawn (i.e. inserted in the display file); if it is not, the object is not drawn.

In Figure 8 the transistor A would be drawn, whereas the capacitor B would not.

There is a significant reduction in the amount of calculation if only straight lines are drawn, while if complex objects such as curves or transistors are drawn, the benefits of this technique are obvious. It also effectively enables an infinite area to be scissored.

There are some disadvantages. More data has to be sorted, and the method breaks down if the object itself is larger than the area that can be scissored. This latter restriction can usually be ignored for all practical purposes.

The objects can be drawn as a subroutine jump to a base point followed by relative visible or invisible vectors. In this way, only the one base point need be changed to reposition the whole of the picture. Even allowing for the fact that some objects may have to be deleted and others added this considerably reduces the number of operations involved in changing the view, a fact that is important whether the display is attached to a small "stand alone" computer or a large time-sharing machine.

Scaling and Zooming

When objects are drawn, it is merely necessary to multiply their co-ordinates and dimensions by a scaling factor to change the size of the picture. In this way, the user can magnify the picture to examine part of it in greater detail. The scale must, of course, be taken into account when performing any scissoring operations.

Other tests can be included when deciding whether or not to display an object, so that, for example, insignificant detail can be ignored. This is not usually necessary but can be essential when representing engineering drawings—when examining a pressure vessel closely, fine detail may have to be shown, whereas if viewing the whole of the chemical plant only the outline shape may be necessary.

Three Dimensions

In just the same way that scaling factors can be applied, objects can be rotated and an orthogonal projection drawn (more complicated views such as perspectives are not usually necessary).

The basic data describing the dimensions and positions of objects should not be changed, but the rotation calculation must be applied to this data anew each time the view is altered. If this is not done, rounding errors will occur and the objects will gradually distort and lose their shape.

Although several views can be displayed side by side, it is seldom necessary to display more than one view at a time as long as the user can easily change from one to the other.

In some cases, it is only necessary to show plan, side, or front elevation thus simplifying the calculation involved.

The user will generally be drawing objects only in the plane of the screen, but the tracking cross can be made to "jump" through the Z-axis when gravity fields are used. The gravity field is applied as if the object were only in two dimensions, but if the cross is repositioned, it is given the appropriate Z-co-ordinate. This reduces the number of view changes needed and makes drawing three dimensional objects much simpler than would otherwise be possible.

Accuracy

It is soon apparent that it is more or less impossible to position the tracking cross to an accuracy greater than about one tenth of an inch. Where the picture has to be drawn accurately to scale, it can often be expanded so that the inaccuracies are insignificant. In other cases, lengths and distances may have to be typed in.

The whole process of drawing accurately can be aided by applying constraints to the process of drawing objects, from the simple ones of keeping lines horizontal or vertical to far more complex constraints beyond the scope of this paper.

Acknowledgements

The following references, in particular, have been found invaluable:

Stotz, R. "Man-Machine Console Facilities for Computer-Aided Design",
    AFIPS proc. Spring Joint Computer Conference, 1963, p. 323.
Sutherland, I. E. "Sketchpad, A Man-Machine Communication System",
    AFIPS proc. Spring Joint Computer Conference, 1963, p. 329.

Biographical Note

A. R. Rundle was a sales engineer with Elliott-Automation Computers Ltd, which is now part of International Computers Ltd (ICL). In 1966 he joined the team developing software for the Elliott 4280 Graphic Display. He is now a Systems Analyst with E–A Systems Ltd. working on display software.

Computer Display System Tradeoffs

HARRY H. POOLE
Raytheon Company

Abstract

This chapter discusses the various tradeoffs which can be made in computer-display systems. Included are eight major areas: (1) display-refresh, (2) category section, (3) word formats, (4) I/O channel capacity and configuration, (5) hardware *v* software, (6) number of displays, (7) display flexibility, and (8) background data. Each of these areas is examined in sufficient detail to indicate the possible alternatives which can be used, the advantages and penalties associated with each alternative, and general guidelines or conclusions where applicable in selecting the optimum approach for a given set of system requirements.

Introduction

The general purpose digital computer and the cathode ray tube (C.R.T.) have been successfully combined in a large number of systems. A C.R.T. display provides an efficient and concise means of disseminating the large quantities of information available from a computer; and in addition, the computer has provided a flexible, efficient means of driving the display. Invariably, however, when two devices of this level of complexity are brought together, a number of compromises must be made. To achieve the proper balance between computer, display, and programming requirements is not easy. There are many factors which must be considered for an optimum system design. This paper discusses a number of the tradeoffs which are available in determining the best computer-display system configuration for a given application. Limitations in the characteristics of selected computer, display, or communications equipment may restrict the availability of some of these choices. Nevertheless, a number of options are usually open in this area, and an understanding of the advantages of each option as it relates not only to the system being built, but to the expansion capability of that system, is also important.

The proper design of any computer display system involves the consideration of many factors. Some of the most important include:

(1) Display Refresh.
(2) Category selection.
(3) Word formats.
(4) I/O channel capacity and configuration.
(5) Hardware *v* Software.
(6) Number of displays.
(7) Display flexibility.

(8) Background data.
(9) Character requirements.
(10) Vector requirements.
(11) Display rotation.
(12) Input processing.
(13) Operator interactions.
(14) Commonality.
(15) Data repeaters.
(16) Data distribution.
(17) Priorities.

The first eight topics will be covered in depth, with an indication given of their interaction with most of the remainder.

Display Refresh

The display-refresh factor has probably the most important impact on computer/display interface design. Display systems should present the required information to the operator for a period long enough to allow him to comprehend and utilise the data. Since his reaction time is measured in seconds, and typical C.R.T. decay times are measured in milliseconds, an interim memory system must be provided to allow continuous presentation of the required data. This memory can be located within the display system through the use of such techniques as display storage tubes, recirculating delay lines, or magnetic core storage. Alternatively, the basic computer memory can be utilised, in which case the information must be transmitted to the displays at rates sufficiently high to provide a flicker free display.

Figure 1 illustrates three typical computer-display configurations, with the two major advantages of each approach also listed. The first system has the computer providing the refresh. In the second system, a separate refresh memory is provided with each display (such as is found in most display terminal applications). The third system provides one common refresh memory at the displays to drive a cluster of displays.

Very often, if the display has its own refresh memory, the computer is forced to maintain a duplicate of this memory file to allow it to update display information or to interpret operator inputs. Therefore, the question to be resolved in a typical system is usually one of input/output capability and limitations, and timing requirements on the computer and the display. The major advantages of utilising a separate display-refresh memory are:

(1) Less memory cycle stealing from the computer.
(2) Lower bandwidth requirements on the computer I/O.
(3) The advantage of a self-contained system for some display operations as well as for test and maintenance purposes.

DISPLAY REFRESH

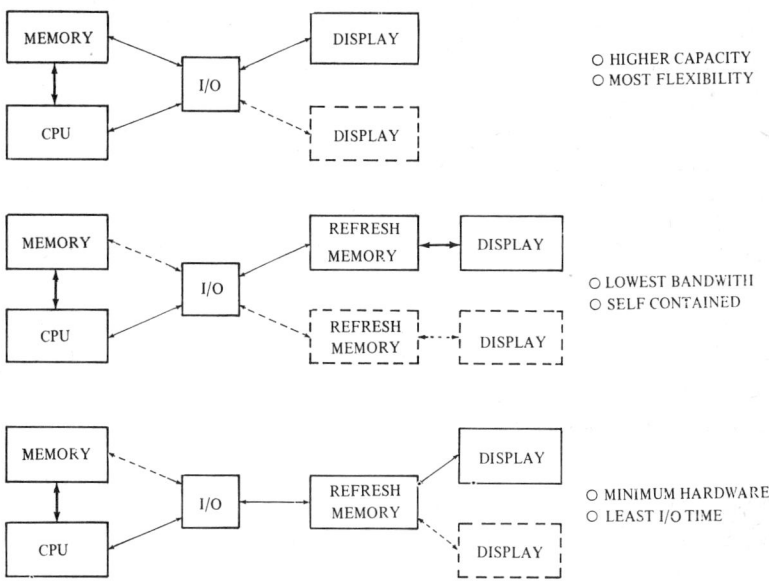

Fig. 1   Three typical computer display configurations

The major disadvantages are:

(1) Added hardware required at each display location.
(2) An adverse effect on display timing and capacity.
(3) Limitations on the flexibility of the display.

Computers which have sufficient capability to provide refresh for a display have direct I/O access to computer memory as well as the capability for simultaneous I/O, and have central processor memory operation through an access bus arrangement and modular memory system. Data being transferred to the display in systems of this type must come from the computer memory, whether for refresh purposes or for updating a memory located in the display. Whenever this data transfer delays an internal operation, available program operating time is reduced. This naturally occurs more often when the computer is used for refresh, since required memory access for refresh typically occurs 10 to 30 times as often as update. This loss of time can be kept to a minimum in systems with a large number of memory modules by keeping operating programs out of the display memory module. Nevertheless, when both the computer and the display require access to the same memory, one of them must wait. This delay is usually assigned to the computer through a technique known as memory cycle stealing. Calculations for typical applications show only a few tenths of 1% loss in operating time per display.

Since the data transfer between the computer and display is typically 10 to 30 times as great if a display memory is not used, it follows that the bandwidth requirements on the I/O channel and the communication link are also greater. This becomes a problem if the distance between the computer and display is large, as high bandwidth communication systems are generally more expensive and less reliable. It is also a problem if the number of displays being serviced by the computer is large, as a separate I/O channel is generally required for every one or two display consoles. With the refresh memory in the display, one I/O channel can handle a large number of consoles. If the computer I/O and communication link have the capability inherent in them for operation at the high speeds necessary to drive all of the displays, this problem will not exist.

The availability of a built-in memory for the display has obvious advantages during production testing and field repair in that the display can be tested as a separate entity. This is not a drawback in most systems using computer refresh since they incorporate a small inexpensive computer as part of the factory test set and use the system computer for determining the source of display failures in the field. However, the self-standing unit also has advantages in certain display operations; since a memory is included in the display with these systems, operator inputs do not impose as high a load on the computer. The display itself can perform such actions as message composition, editing, cursor control, reformatting, display expansion (to name a few), without the necessity of interrupting the system computer to insert data and have this data acted on. Although the display has a low interrupt rate to the computer, saving on the number of interrupts which must be processed still increases available program operation time for other functions.

The incorporation of a memory within the display generally increases system cost, size, and weight, and decreases system reliability. This lack of reliability occurs because when the display memory fails, the display can no longer be used. However, if one of the computing memory modules fails, a spare module can be used in its place, resulting in negligible effect on system operation. An increase in the number of special purpose logic and control circuits is also required with a separate display memory, especially in multi display systems.

An effect may also be felt on display load capacity. In most systems of this type, the display refresh function is interrupted to accept new, updated information from the computer, providing less time to read the data out of the memory. Display systems seldom have the flexible logic and timing necessary to interleave the functions of read and write to reduce this loss in time. Since the display logic is hard wired, there is also an effect on display flexibility and expansion capability, as this type of logic can never be as flexible as programming. Various possible conflicts or overload situations must also be solved by hardwired means, which limits the flexibility of the system in these areas.

In conclusion, it is safe to say that if both the computer and communication systems have the necessary capability, it is usually less expensive, more reliable, and more flexible to have the computer refresh the display. For limited capability computers, multi-display systems, or long communication paths, it is more desirable to have the refresh memory in the display.

Display Category Selection

The second factor to be examined is display category selection. When the operator requests new types of data, the implementation of this request can be performed through computer programming or through display hardware. If the display performs the selection, the computer is required to send all of the available data to each display. However, if the computer performs the selection, only the desired data must be transmitted, with separate data being transmitted for each display. The choice, obviously, has an impact on both computer programming and display hardware. It also affects a number of other areas such as data word format and length, display load capacity and timing, reliability, and display flexibility. Although it has only limited impact on the refresh memory location, it does have an effect on its size.

Figure 2 illustrates the two configurations for performing category selection with the refresh memory in the computer. Similar approaches could be used if the refresh memory was located in the display. In the first configuration, all of the data is sent to the display system and the

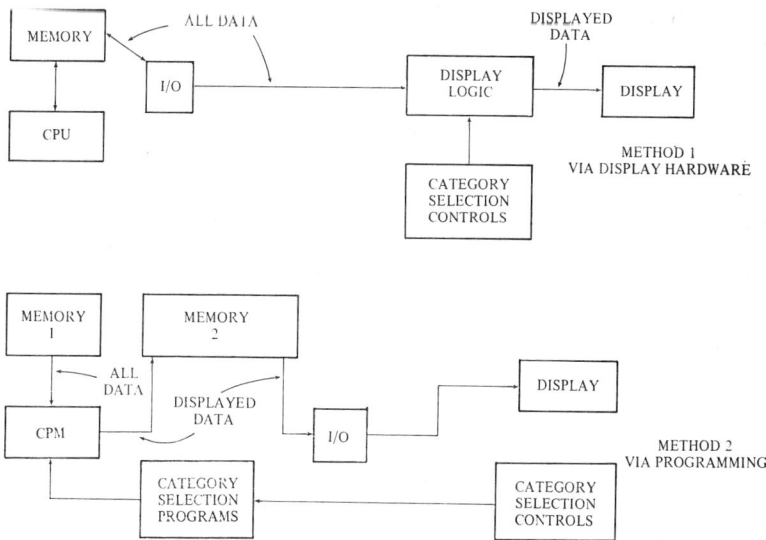

Fig. 2 Two configurations for category selection with computer refresh

display logic selects that portion to be displayed. In the second configuration, special category select programs are used in the computer to limit the information being sent to the displays to those categories selected by the operator.

Category selection buttons are used by the operator to select the classes of data he will see. Actual data classes depend on the function that the display is called upon to perform; for example, in an air traffic control application, they may include such items as: controlled aircraft, all aircraft, air corridors, map data, weather, and alternate airport locations. In addition to the information under operator control, certain other data may be forced on the display. Included in this category would be information of a critical or emergency nature, such as an impending collision.

For the display to provide a category selection capability, category information would have to be transmitted with the data. This can be done through the use of mode identification bits associated with each data word. Alternatively, it is possible to send out all data of one class in sequence, thus requiring only a mode control word to precede each class of data. In practice, this latter approach imposes programming difficulties by requiring all the data in the memory to be re-ordered whenever new or different data is added to the memory. In either case, selection logic in the display would determine the data to be presented on the display by examining the mode bits associated with that data.

This approach involves difficulties both in implementation and in display capacity. An appreciable amount of additional hardware is required for the display to perform category selection, since one set of logic decoders and one buffer register is required per display category. However, the computer requires no additional hardware if it provides the category selection function. As regards capacity, an amount of time is now required to look at each word to see if it should be displayed, thus reducing the available time for displaying the data.

The additional mode bits required for the data words increase the total number of data bits required, thus increasing the size of the refresh memory. In addition, since all possible data must be in the refresh memory, rather than just the selected data, the size of the refresh memory is further increased, becoming so large in some cases as to be prohibitive. Finally, there are often a number of data words which belong in more than one class. These data words pose coding problems if display selection is utilised. In summary, this approach has a major impact on many areas of the system; a sizeable increase is required for both display data memory and instruction memory, I/O channel data rate usage is greatly increased, and display capacity is reduced.

If the computer performs category selection, however, there could be an impact on programming requirements. For the computer program to perform data category selection, special subroutines must be called up every time the operator changes a category selection button. These subroutines are used to determine the specific addition and removal of data from the display refresh memory. A data search must be conducted to determine what data to add, and in turn this data must be properly formatted. To accomplish this, computer time will be required in the order of a few milliseconds for a typical category request. These category selection sub-routines will be called up relatively infrequently, since the operator must have time to analyse the information which he has just received before he will request new data.

In a typical operational situation, category selection requests may be made at rates as high as about ten requests every few seconds; however, more normal button activation rates would be in the order of 1 every few minutes. This means that approximately 1% of the available computer time is required during peak conditions with only 0·02% under more normal conditions. In either case, this does not constitute an excessive computer load. It should be pointed out that some of this time would have been utilised even if the display provided category selection, since new information which will not be displayed must still be updated. Since update frequency is greater than selection frequency, it may actually save machine time by having the computer perform this operation. Program formatting is also slightly simplified in this approach. In summary, the difference in operating program time between the two approaches is probably negligible for most applications.

Other factors should also be considered. System reliability is improved if the computer performs category selection, because less hardware is required. Identical displays can now be utilised for different types of operations since different category selection logic is not required for each console type. As mentioned previously, display timing is affected if display selection is implemented. Each display must receive all data words available, rather than just those to be displayed. This requires a greater portion of the available display cycle time to perform this added function regardless of where the refresh memory is located. The additional time required would depend on system application but would typically run from 2 to 10 microseconds per unwanted item. Since there may be several hundred unwanted items stored in the refresh file, an appreciable percentage of the available display cycle time could be required.

There is yet another factor to consider. When one common refresh memory is used to refresh several displays, the situation changes. Whether this memory is located in the computer or at the display subsystem, when it services a number of displays the size of the memory will be reduced if it holds one file of all the data rather than several files of part of the data. Computer time to generate these files will also be reduced. In this case, the question of which approach is best may hinge on the amount of data which is common to all displays and the desired degree of flexibility within each display.

In conclusion, if there are a large number of displays, each with their own memory, category selection by the computer is the best approach.

If there are a number of displays sharing the same memory, the decision would be based on how much data was common between the displays and what was the maximum size of the data base. With a large data base, and a large amount of data common to all displays, computer selection would be preferred; while if both were small, display category selection would be required. In systems having few displays, computer category selection is generally the best approach.

Word Formats

Transfers between the computer and the display may be made serially, parallel by character or parallel by word. Serial transmissions are usually utilised in those applications which have the computer and the display widely separated, such as airline reservation systems. Parallel transfers are employed when the display and computer are co-located. Character transfers are used with the more limited displays, such as electronic typewriters or tabular displays, where the only information transmitted is the character itself, with position being defined according to predetermined formats. Word transfers are employed with more complex graphical or situation displays, where vectors as well as characters must be displayed and several types of word formats, some containing major position information, are needed. These three types of transmission are illustrated in Figure 3.

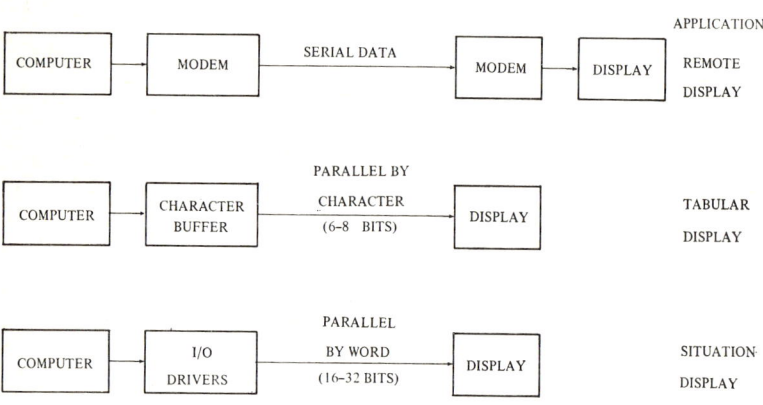

Fig. 3   Three types of word format transmission

With serial data transfer, the refresh memory usually has to be in the display, as the channel capacity is too low for computer refresh. Category selection is also generally done in the computer, for a similar reason. For this type of transmission, word formats are usually dictated by the requirements of the communication system or the data modems employed and not by the computer or the display. One exception to this occurs when error detecting or error correcting codes are employed. In most cases, a simple parity bit is all that is used because of the extra circuitry which is required for some of the sophisticated error correction codes. In these applications, the operator is responsible for detecting multiple errors. Upon the detection of an error, retransmission of data is requested.

Another type of error detection code which is employed for serial transmission between the computer and the display is the use of a known format. This known format may be simple, such as the teletype code, the more optimum ARQ (Automatic Repeat reQuest), or a 4 out of 8 code. Error correction codes are seldom advisable for computer-display data transfer because they require more bits to be transmitted, can detect fewer errors, occasionally add errors, and require extensive hardware to implement.

Error detection codes can be employed in systems using parallel data transfers. For example, parity bits are usually added to detect equipment failures in these systems. However, it is only lengthy communication lines, which invariably employ serial transmission, that are susceptible enough to noise, fading or crosstalk to warrant the use of more sophisticated codes.

When data is transferred by character, the application is usually one in which only tabular data (i.e. characters arranged in a predetermined format) is to be displayed. The only choice to make is whether to employ six-, seven- or eight-bit characters, and with or without a parity bit. With parallel data transfers, at least six bits are always used to define a character. Although 64 character codes are usually sufficient for most display applications, there has been a growing attempt in the United States to standardise on a character code sponsored by the American Standards Association. This code utilises seven bits, although a six-bit subset can also be utilised. By having 128 character combinations, this code allows the use of a large number of control characters as well as the basic alphanumerics, thus making it useful for line printers and typewriters as well as C.R.T. displays.

The use of eight data bits to define a character allows the basic seven-bit character to fit into the 24-, 32- or 48-bit computer word more readily. It also allows a more extensive error detection code (such as 4 out of 8) to be employed for serial data transmission. Finally, it is more compatible with some magnetic tape units.

The system designer has a great amount of flexibility when it comes to word data transfers between the computer and the display. It is also the time when the programmers and hardware designers will have a number of good suggestions to offer. The number of bits in the word is usually fixed by the choice of computer. From a display point of view, 24 to 32 bits would be optimum, depending on the number of control bits required. Displays can, however, interface with any length computer word, although the word length may lead to an inefficient data transfer in some applications.

E

The information which is to be transferred on the computer display interface includes data defining the display mode, the type of character or vector to be displayed, the position at which it is to be displayed, and often a number of special modifiers. These may include character size, brightness, rotation and blink as well as vector brightness and blink. In addition, if several characters are to be positioned in a given format (such as horizontal spacing), the number of such characters must often be specified.

All of these factors give a number of options to the system designer. He could have all words of a given mode transferred together in a block, with a mode change word inserted between blocks. This not only removes the necessity of having mode bits inserted in every word, it also often simplifies the display control logic and timing. Unfortunately, as discussed earlier, it does this at the expense of the computer programmer who now must arrange all the data into the proper order. It will also have an adverse effect on display capacity since characters and vectors written at the same location on the C.R.T. require two major positioning commands instead of one.

If mode bits are inserted in every word, they are usually placed at one end of the data word (most commonly the least significant bits) to aid the programmer. The number of mode bits is usually one or two. One bit would be used to distinguish characters from vectors, while two bits will allow distinctions to be made between formatted and between single and continuing vectors.

The number of bits associated with position information is usually from 16 to 22, with 18 and 20 the most common. This allows 9- or 10-bit accuracy for both X and Y coordinates, which makes this accuracy approach the basic resolution of the C.R.T.

Variable size characters (in two or four steps) are often employed in display systems, with variable brightness characters (again in two or four steps) less popular. Character rotation (90°) is useful in applications requiring charts or tables, and blink is advantageous as an attention getter in most systems. Different blink rates (2 or 3) are also occasionally used.

In the vector area, there is often a need for dashed or dotted vectors as well as two brightness levels. Vectors can be drawn based upon end point data (the most common), incremental data, or slope and magnitude (the least common).

Defining a typical data word is very difficult, as both the number of bits in the word and the information to be coded into the word vary with application. However, a typical 32-bit word may be configured as shown opposite.

In conclusion, word formats in serial transmissions are primarily dictated by the requirements of the communications link and the need to use special codes for error reduction. Parallel word transfers offer a great amount of flexibility, within the constraints imposed by the computer word length and the number of special control bits required.

Computer Display System Tradeoffs

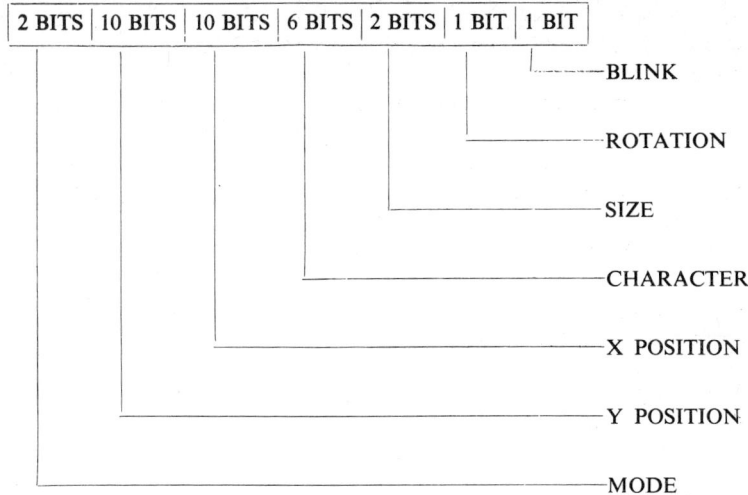

In the worst case, it may take two word transfers to convey all the information required for one data point.

I/O Channel Capacity and Configuration

The computer input/output channel characteristics have an impact on the display system word length, required number of control signals, timing, and display capacity. Most computers have the capability of providing word parallel output, as well as providing data in bit serial form through a communications buffer. In addition, some computers such as the SDS 920, contain character buffer outputs, which allow data to be transmitted in six- or eight-bit groups. For most computer graphics systems, the choice is usually between serial data transfer through a communication system or parallel data word transfers.

Several control line approaches can be used. Separate data lines for input and output allow a simplification in the display control logic which governs input/output traffic, but require more data lines and associated hardware. Unless these lines can be utilized for simultaneous transfer of data in both directions, little traffic simplification has been achieved. Without simultaneous transfer, usually requiring two computer I/O channels per display system, some buffer storage must be provided to store temporarily input data from the display.

The actual interface data transfers can be either pulsed or through a level change with sampling of the data lines. In the first case, control signals are required before the data exchange to indicate that one of the devices has data to be transferred and that the other is ready to accept data when it is put on the lines. Simultaneously with the line being pulsed, a third control line is required to indicate that fact, while the receiving device activates a fourth line to indicate that the data was successfully received.

Because of the precise timing required with pulse data transfers, a number of systems employ level changes. In this case, information remains on the line until accepted by the receiving end or until a predetermined amount of time passes. This latter factor is used to prevent hang ups in the case of equipment malfunction.

In either approach, separate control lines must be provided for data transfers in both directions. These include the functions of data ready, data acknowledge, parity error, line busy, and for timing control. They are concerned only with controlling the flow of information between display and computer. The information so transferred is then interpreted by the display itself as display data (e.g. for locating a new spot) or display control (e.g. for setting up the "blink" condition, as described in the previous section).

Display data capacity will be adversely affected in systems employing computer refresh if the display must wait for service between each data word. If the average wait is less than the time that it takes the display to utilize an average word, this delay can be minimised by double buffering.

Hardware v Software

In addition to the factors discussed above, the hardware-software tradeoff affects most areas of display implementation. Display rotation, display offcentering and expansion, vector generation, and formatted displays are typical of the functions which can be performed with or without program intervention. Each of these areas will be discussed below. However, a few general comments relevant to the hardware-software tradeoff should first be made.

In general, any of the above operations can be accomplished by either programming techniques or through the use of hardware. Those which are difficult by one approach are usually difficult in the other as well. The choice as to which ones should be hardware implemented and which ones software implemented should be based on such factors as the number of displays being serviced, the frequency with which the item in question will be requested, other features of the display which are already hardware implemented, and the expansion potential of the system. System expansion generally dictates a software implementation, as it is difficult to add hardware after the system is built. For this reason, as well as for contingencies that develop during system test, there should always be left enough programming time and flexibility in the system to allow system expansion, and therefore, where practical, it is probably best to implement as much as possible via hardware. The limiting factor of course, is that there are functions which may be too difficult to implement with hardware.

First consider display rotation. In this case, the entire display presentation is rotated about the Z axis upon command, causing the X and Y axis to be interchanged after 90 degrees of rotation. This rotation is relatively easy to make via hardware, while requiring a fair amount of computer time if software is implemented. There are, however, display graphics applications which require rotation about every axis. In this case, three sets of coordinates have to be known and changed for each data point. Implementing this full rotation option through hardware would require extensive arithmetic operations within the display system. For that reason, rotation about any axis must usually be performed by the main computer, under program control.

Display expansion can be done readily via hardware but problems develop when this is coupled with offcentering. For both to be accomplished via hardware requires the addition of a limited arithmetic unit, if the operation is to be performed digitally. If analog offset is employed. some inaccuracies due to limitations in the deflection system will result. When both are performed through software, the resultant computations (shift and add) are relatively simple, having to be performed on each data word only whenever a scale change or offset is desired. If these operations are done in the display they would be required for each refresh cycle. For this particular area, it should be concluded that offset and expansion can be done in either location, with the choice dependent on the number of displays in the system (thus affecting the amount of hardware which must be built) and the total load that this function would impose on the computer.

Formatting the displays conserves program time as well as the number of words in the refresh memory. One example is a tabular display. This device presents all information in typewriter fashion, i.e., each character is placed horizontally to the right of the last character, with line advance and reset automatic once the last character is typed on a line. Since no major position information has to be computed or stored for any character, there is a large saving in both memory and operation time. Even graphical displays very often have to present information that can be formatted. This feature can be implemented with a limited amount of hardware, especially if automatic line advance and reset is not required (which is usually the case for graphics displays). The format can be horizontal, vertical, or a combination of the two. All that is required is that the display include the necessary hardware and that a special mode word or bit define the data which is to be handled in this manner. It is possible to combine a vector with several characters in a format. Therefore, if a given display arrangement is to be used repeatedly, it will usually prove more profitable to perform the formatting via hardware, and save the computer time for some of these other functions.

Even the vector generator can often use some help from the computer. Depending on the type of vector generator employed, it is usually necessary to know not only the change in both the X and Y axes, but also the vector sum of the two. The latter is needed for both vector straightness and brightness uniformity (the computer can provide both sets of information). However, a limited amount of display hardware can be used to provide an approximation of the vector sum, to a sufficient accuracy for the purpose, and without imposing an additional load on the computer.

In conclusion, most systems should utilise display hardware solutions to these display functions if at all practical, with extensive display rotation being the only clear-cut example that should be performed within the computer.

Number of Displays

Many computer-display systems have a large number of displays driven by the same computer. This factor has several impacts on the system configuration, particularly in areas of commonality, category selection, and refresh location. To minimise design, production and maintenance costs, it is highly desirable to have multiple displays as nearly alike as possible. This means that many of the displays will have added features or circuitry that are not needed for their specific use, and it may mean that the design of some has been compromised for the requirements of the many. In any case, the benefits of display commonality are usually great enough to pay for the limited amount of extra hardware that may be required. If the requirements imposed on the different units are too diverse, it may still be possible to have a basic display unit with various plug-in modules to provide the desired diversity.

Computer refresh can be used with multiple display systems, if the number of displays required is small, and the distance to the central computer is short. In most cases, however, the refresh function is moved into the display equipment in multiple display systems. With the latter approach, a common display refresh memory can then be considered for groups of displays that are co-located, or each display can include its own memory. The major advantages in having one common display refresh memory include a reduction in total display storage required and the inclusion of the category selection function in the display, thus saving computer time. This is practical as long as the total memory file that a given display can request data from is not too large, and as long as there is common information to be presented on several displays.

In operation, the common memory approach would cycle the memory through all of the words in storage, with category selection logic from each display determining which portions of the refresh file it should receive. The memory would step to the next word automatically after a fixed time delay, which would be sufficient for the type of data to be presented. In this approach, all of the data stored in the refresh memory could be written on any display with the average display presenting perhaps one-third to one-half of the data. By using a double buffer system, and more complicated memory control logic, it is possible to make more efficient use of the available time, thus increasing the data base available, while not increasing the maximum amount which can be shown on any one display.

It should be noted that a common display refresh memory and independent category selection will also allow a common character and vector generation system to be employed, thus providing a further saving on hardware. In fact, it is possible for the display system to be comprised of multiple analog displays (including the C.R.T., deflection amplifier, video amplifier, and power supplies) with but one common interface and control area required. This common area would include the computer interface, refresh memory, category selection logic, character generator, and vector generator.

Figure 4 illustrates this type of multiple display system. A common input/output area, refresh memory, character generator, vector generator and category selection logic are provided, with multiple display terminals containing only the basic analog portion.

MULTIPLE DISPLAY SYSTEM

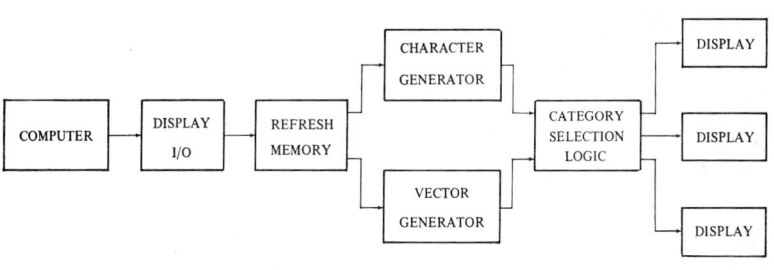

Fig. 4

In conclusion, if there are a large number of displays in the same area to be driven by the computer, a new approach to the display system can be taken. This approach would timeshare as much hardware as possible within one common area, thus saving appreciably in overall hardware costs. It also allows a slightly different approach from the single display system to be taken in the areas of category selection and refresh. If the multiple displays are widely separated, communication criteria require them to be independent and autonomous.

Display Flexibility

A discussion of the impact of several approaches on display flexibility and expansion was given under earlier topics. It was mentioned that performing various functions through software provided in general the most flexible display, since the programs can be rewritten to accommodate new or changing requirements. There are also other features which can be implemented within the displays that have an impact on the computer-display system. This section will discuss the general usefulness of a number of these display options.

Blink is an excellent attention getting device. It is even practical to employ two blink rates, a high rate (6 to 10 cps) for critical attention getting, and a low rate (1 to 3 cps) as a reminder or feedback technique. Multiple blinking targets can prove very distractive, however, so that it should be limited to only a few targets at most.

Circle generators, especially when they will also produce ellipses, are very useful in a number of display graphics applications, such as computer aided design. It is of less use in other types of displays, such as tabular or situation displays.

Dashed and dotted vectors are a worthwhile option in any graphics display system and can be added easily to most vector generators that write at a constant rate. For some reason, most current systems do not take advantage of this capability.

Character rotation has been supplied on a few systems. A rotation of 90 degrees is not difficult but has only a limited application, such as the display of tables. Unlimited character rotation requires extensive hardware and for most systems is not feasible or warranted.

Presenting characters in several sizes is a popular feature in a number of systems. There is a limit to the number of useful sizes which can be presented, since very small characters cannot be seen and large ones take up too much space, thus cluttering the display. For this reason only two or four sizes are supplied. The variable size can also pose a problem if used in a formatted display, as variable spacing will be required as a function of character size.

Characters and vectors of variable brightness is another option. For most applications, all vectors should be seen as background to the characters on the screen, therefore, it is unlikely that more than two brightness levels are needed for either characters or vectors. The brightest vector should have the same intensity as the dimmest character. It should also be noted that in practice, most display systems having four levels of character brightness cannot be kept in adjustment to provide any more than three levels.

DISPLAY FEATURES

FEATURE	APPLICATION
BLINK	ATTENTION GETTING
CIRCLE GENERATION	DISPLAY GRAPHICS
DASHED / DOTTED VECTORS	DISPLAY GRAPHICS
CHARACTER ROTATION	CHARTS / TABLES
MULTIPLE CHARACTER SIZES	EMPHASIS
VARIABLE BRIGHTNESS	BACKGROUND DATA

Fig. 5

This section has briefly described a number of display options which are often of importance in computer driven display systems.

Figure 5 lists six such features, together with typical applications in which they could be used.

Background Data

One final area should be covered that (although somewhat different from those previously discussed) is of importance in a number of computer display systems. This area is the presentation of background data. This data includes such items as geographical map data for situation displays, as well as charts and tables for various business applications. There are four basic ways in which this material can be presented on the display: overlays, optical super-imposition, electronic mixing, and computer generation. Since it is seldom practical to change overlays except on an occasional basis, the overlay is restricted to applications which do not require a change in background data such as some medical electronic applications. Overlays also have the capability of being edge-lighted, thus allowing the operator control over the intensity of the background data.

Optical superimposition is a second major class of techniques for providing background data. Three types are typically used: front projection onto the cathode ray tube screen, rear projection through a ported C.R.T., and the projection of both the C.R.T. image and the background image. In most cases, a slide projector containing from 20 to 100 background slides is used. With front projection systems, the slide projector may be either placed over the shoulder of the operator, or above the console, with the image projected onto a C.R.T. screen at a 45 degree angle.

In rear projection systems, a ported tube is used. This is a C.R.T. which has an optical window on one side to allow the projector to focus through the window onto the screen. The tube must be carefully designed to prevent both optical distortion of the slide image and electrical distortion of the C.R.T. image caused by the nonhomogeneity of the bulb.

Optical superimposition systems have the advantage of producing the background data in color, providing quick and often automatic slide change as required by the system, and producing images with more complexity than is possible through either computer generation or electronic mixing. They produce images of medium registration, typically 1 percent. Their primary disadvantages are the optical distortion (usually "keystone") present, and the fact that each console requires its own background data system. Figure 6 illustrates the three types of optical superimposition.

Electronic mixing combines the video signal produced by a video mapper or monoscope tube with the desired character and vector information by timesharing the C.R.T. beam between both sources of data. Unless a dual gun tube is utilised, this method cuts down on the

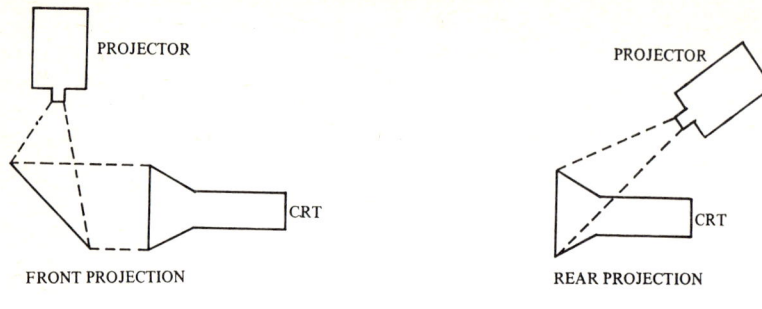

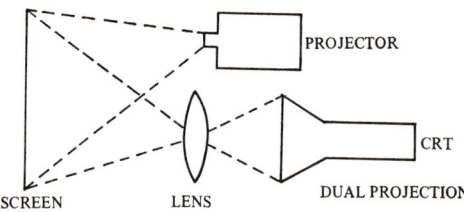

Fig. 6 Three types of optical superimposition of background data

information which could otherwise be presented. It also produces images of poorer resolution than any other method. Its major advantages are that it can produce backgrounds of greater complexity than computer generated backgrounds, without requiring computer time, and can produce the desired background data for use on a number of displays at one time.

In graphic display systems, most background data is generated by the computer and put on the screen via the vector generator. By using this approach, it allows direct operator interaction with the computer, by allowing him to enter and alter or delete lines as required. However, there are whole classes of systems that do not require this operator interaction but use a few standardised forms at many locations. These systems could use other background data techniques, thus saving computer operation time and refresh memory space. Two examples of this are in airline reservations systems and hospital systems. In each of these, the form is for the convenience of the operator only and generally is not altered once installed. In this type of system, serious consideration should be given to a centralised background method, such as the video mapper or the use of local overlays.

Conclusion

This paper has discussed eight tradeoff areas to be considered in the design and development of computer-display systems. Following each discussion, a summary was then presented. Since the items covered were tradeoff areas, no firm rules could be given for determining which approach is best in a given application. Instead, the relative advantages and disadvantages of each option were discussed as well as their applicability to different system configurations. Because of the interaction of each tradeoff area with others, there is usually no approach that is clearly optimum, and the system designer thus has a great deal of flexibility in deciding which parameters are the most important for his application.

Biographical Note

H. Poole is Principal Engineer in the Missile Systems Division of the Raytheon Company, Bedford, Massachusetts.

Computer Graphics in the United States

C. MACHOVER
Information Displays Inc.

INTRODUCTION

In the United States, cathode ray tube graphic terminals have been associated with digital computers for the past decade or so. Initially, the terminals were used in military command and control systems. For example, in the mid-fifties, the SAGE System (an air defense system) used C.R.T. terminals which were not grossly different from present-day units.

However, the use of C.R.T. graphic terminals in a non-military environment is relatively new. The digital system historian, if there were such a discipline, would probably date this application of C.R.T. terminals to the pioneering work done by Dr. Ivan Sutherland on Sketchpad[1] in the early 1960's. My company, for example, has been involved in computer controlled display systems since 1960 and, I must admit, that for several of the early years we felt that we had a cure for which there was as yet no known disease.

Today, however, the use of graphic terminals in both research and profit-making environments is growing rapidly. The Brunel Symposium, for example, was typical of those which have blossomed out during the past few years in the United States to reflect the growing interest in graphic terminals. Professional societies, such as the Society for Information Display and UAIDE, have been formed to service the interest of workers and users in the field. The older professional societies, such as ACM and IEEE, have recognised the importance of graphic terminals by creating sub-groups whose primary area of interest is these devices. All of these are surface indications of the growing importance of this tool. One authority estimated that in five years about 13 cents out of each computer dollar would be spent for C.R.T. terminals.

How many C.R.T. graphic consoles* are now installed, or are on order? How does this compare with the situation five years ago? Frankly, I don't know. I have not been able to find reliable statistics. Based on the information I do have, however, I would take an educated guess that there are probably 500 to 1000 graphic consoles now installed or on order. In 1964 there were probably no more than 100. If the authority cited earlier is correct, there will be a billion dollars worth of terminals bought in 1973. Assuming that the average price is $50,000, 20,000 terminals will be sold that year. It is an interesting number but I cannot vouch for the accuracy. In any case, the present is bright, and the future is exciting.

* Mr. Machover means interactive vector devices and not alphanumeric terminals throughout his chapter—*Editors*.

In this chapter, I will: review some of the applications in which the C.R.T. graphic terminal is currently being used and where it appears to hold promise for future use; describe some of the commercially available U.S. hardware; prognosticate a bit on future developments; and finally, make some comments about the software aspects of C.R.T. graphic displays.

APPLICATIONS

How is computer graphics being used commercially in the United States today? In this section, I will discuss applications in the following categories:

Computer Aided Design
Management Information
Simulation
Process Control
Computer Aided Education
Pattern Recognition
Graphic Arts
Computer Generated Movies
Others

Computer Aided Design

In the areas of computer aided design, there appear to be several developing applications. Mechanical design (including numerical control) is exemplified by the work of Lockheed-Georgia and General Motors.

Aircraft Industry

Lockheed's effort in the area of mechanical design using graphics is indicated in a recent trade press report:[2]

"Interactive graphic displays ... are being used by Lockheed-Georgia Company, a division of Lockheed Aircraft Corporation, for numerical control part programming of continuous path machines. The procedure replaces A.P.T. (automatic program tools) programming system and makes it possible to produce a verified part program in as little as one hour. The system is being used to produce C5 parts and tooling jigs. 'We have programmed about 50 parts for the C5', says tooling supervisor, C. F. Nicks, 'and more than 300 tooling jigs'. By drawing the cutter path over the part blueprint displayed on a C.R.T., the designer tells the computer to set up a proper program to cause that particular action. The computer then displays the path it has programmed so that the designer can see that it is correct."

Indications are that other aircraft companies, particularly McDonnel-Douglas and Boeing, are also using displays in this way.

Automotive Industry

Outside of the aircraft industry, perhaps the best-known program of computer aided design using graphic terminals has been mounted by General Motors with DAC (Design Augmented by Computer). DAC has been an active program with General Motors since 1962 and the work has strongly paralleled the original concepts of the Sketchpad.

In describing the justification for the DAC system, E. L. Jacks of General Motors, asks:[3]

> "When you consider both man's and computer's time to do a job, is such a time sharing system any faster than the old method of processing computer jobs one at a time? From the user's point of view, is the random time shared approach more efficient than the carefully pre-planned 'batch-processing system'? For both questions, the answer is a definite yes; it may be eight to ten times faster using time shared consoles based on actual operating experience with the G.M. DAC-1."

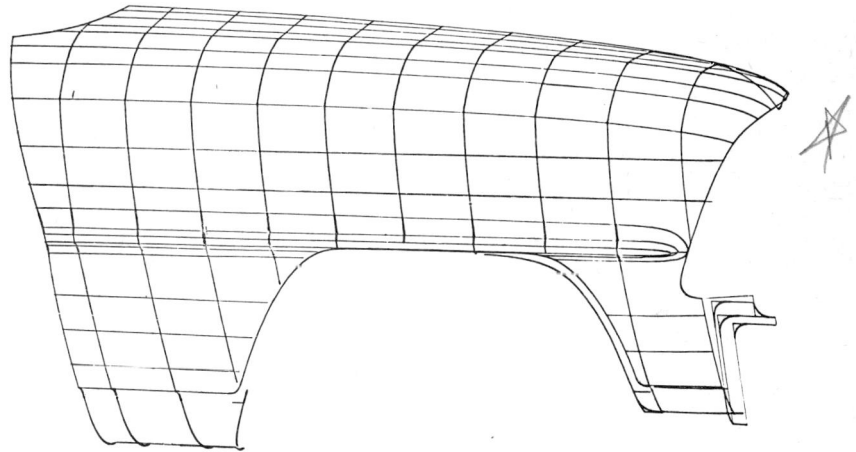

Fig. 1   Typical drawing produced by General Motors DAC System

Indications are that General Motors are now using these systems in the design of their current automobiles.

Figure 1 is a typical drawing made with the DAC system.

There is some indication that other automobile companies are beginning research in these areas; apparently Ford Motor, for example, has developed programs for designing windshields.

Integrated Circuits

One other specific area of computer aided design into which graphic consoles are making significant inroads, is the design of integrated circuits. Much of this work was pioneered by the Nordern Division of United Aircraft working in conjunction with IBM. Motorola's effort in this area have been reported in the trade journals.[4] Other integrated circuit manufacturers who have begun to show an active interest in the use of Cathode Ray Tube terminals for integrated circuits include Bell, R.C.A., G.E., Fairchild Semi-Conductor and Texas Instrument.

Textile Design

Not all computer aided design projects are limited to the drab world of manufacturing. Perhaps one measure of the increasing use of computer graphics was the inclusion in the IBM Pavilion in Hemis-Fair 68 (San Antonio, Texas) of an IBM 2250 linked to a 360-30.[5] During this demonstration, a computer system was used to weave a cloth in designs done by visitors. The visitor uses a light pen to draw a design on the display screen and indicates what types of weaves he wants used in different parts of the design. The information is then translated into instructions for an 11-foot high loom. Coloured threads are automatically introduced into the weaves by the loom. The output of the system is a 3-inch square swatch of fabric for the person who designed it. In the application, the designer is able to select a weave from a library stored in the computer and have it inserted in all

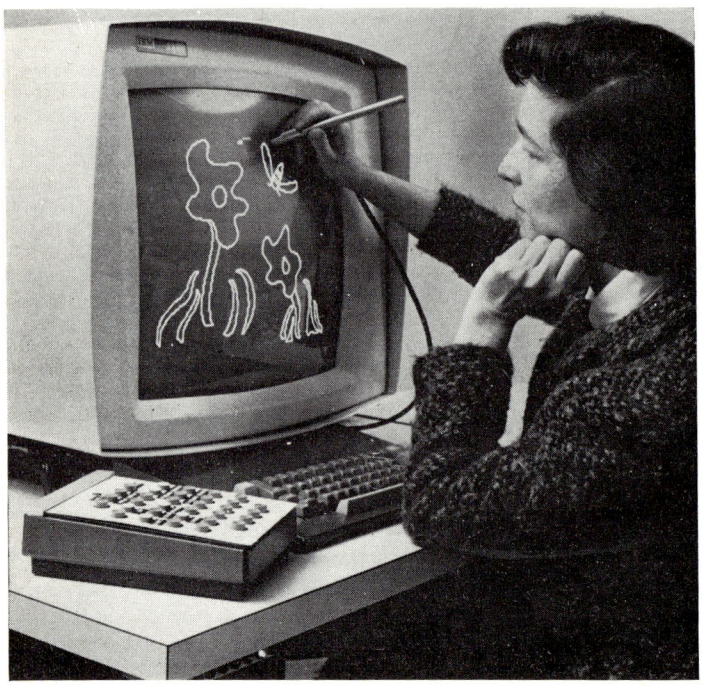

Fig. 2   Textile design with graphic console (by permission of IBM)

appropriate areas of the design simultaneously. Figure 2 shows a textile designer at work.

Conclusions

Certainly many of these computer aided design applications are still in the experimental stage. A U.S. Government report[6] commented:

> "All of the installations currently using this type of equipment are classed as experimental.... Graphic consoles ... are still several years away from general operational usage."

If we were to modify that statement as it might apply now, three years later, we could say that *most* of the installations currently using this type of equipment are classed as experimental. However, in selected applications, the past experimenting was completed successfully and the equipment is now being used in a profit-making environment. Further, we can fairly say that graphic consoles are *beginning* to come into general operational usage.

Current users claim reductions of from 2 to 1, to 6 to 1 in the time required to perform certain design functions. There is more than sufficient justification for management interest in this type of increased productivity.[7]

Management Information Systems

The complexity of the job to be done by the modern manager has led to great interest in applying computer graphics. Westinghouse now uses an on-line system (consisting of a UNIVAC 494 and a IDI display) for sales forecasting and production planning of their washing machine line. It is reported that Boeing has installed a graphic management information system which allows such things as on-line PERTing. Boeing also installed a similar system for the Air Force, and System Development Corporation has implemented an on-line system in a time sharing environment.

It will be instructive, I think, to examine in some detail the manner in which the Westinghouse and SDC systems have been implemented. These demonstrate some of the characteristic features of interactive displays in this type of environment.

Primarily, the display is used as a "what if" device. It allows the manager to ask the question "what if?" in the context of changing inventory levels, changing production levels, and a variety of other things which are so much a part of the management environment. In this environment, the problem is to create a tool with which the non-computer specialist can work comfortably. The tool requires presentation of the problem in the non-specialist's language and the translation of non-specialist's decisions and questions (made in a form most convenient for him), back into computer language.

In the Westinghouse application, for example, all parameters of the problem are initially presented to the manager in a form of a light

F

pen menu. By pointing the light pen at the information on the C.R.T., the manager can choose a period over which he wants the information displayed, can choose the item for which he wants the data shown, and can specify the form in which the information is to be presented. The displayed data can be manipulated by light pen and keyboard. In this way, the manager conveniently moves back and forth within data base, asking "what if?", and making decisions. Although the data is computer generated, it is presented in the form with which the manager is intimately familiar. He need not learn a new language in order to communicate with the computer. And his response from the computer is fast enough so that he doesn't forget his original "what if" question.

The System Development Corporation in Santa Monica, California, has been developing a display system over the last several years for use in a time sharing environment. Their system, using a prototype program called DISPLAY, is designed to provide automatically determined standard graphic presentation in response to the user's light pen inputs. An underlying principle in the design of the display was to make the scope face serve as a helpful guide to the user in creating a display.

For example, to achieve a simple scatter plot, like that shown in Figure 3, the user provides the DISPLAY with five light pen actions. It takes two inputs to specify the X variable, two to specify the Y variable, and one to cause the program to exit. Axis scaling, axis labelling, data scaling, data plotting, are all performed by the program. The user may browse through his data base under light pen control. Figure 3 illustrates some of the flexibility of the display. The picture was achieved by moving through a series of operations under light pen control in a prior display. Titling, axis labelling and graduation, as well as the scaling of the data, where automatically performed by the program after the title and labels were selected. The control button at the bottom of the scope allows the user to read out the value at a specific point. The notices at the bottom of the picture are the exact values of assessed valuation and taxes at the point located by the cross. The user may also delete data labels or titles. The light button AXIS SCALE permits the user to re-scale his data to examine some area of particular interest to him. Other light buttons include: SAVE (permits the user to put a picture and data away on peripheral storage for subsequent examination); and START-OVER (the escape hatch necessary to recover from unfortunate use of the various touch-up options).

Simulation

Simulators are using C.R.T. terminals as the output device and have become fairly widespread. Included in the pioneer work is the SKETCHPAD program which allowed simulation of a bridge design. The structure was drawn on the face of the C.R.T. and calculated.

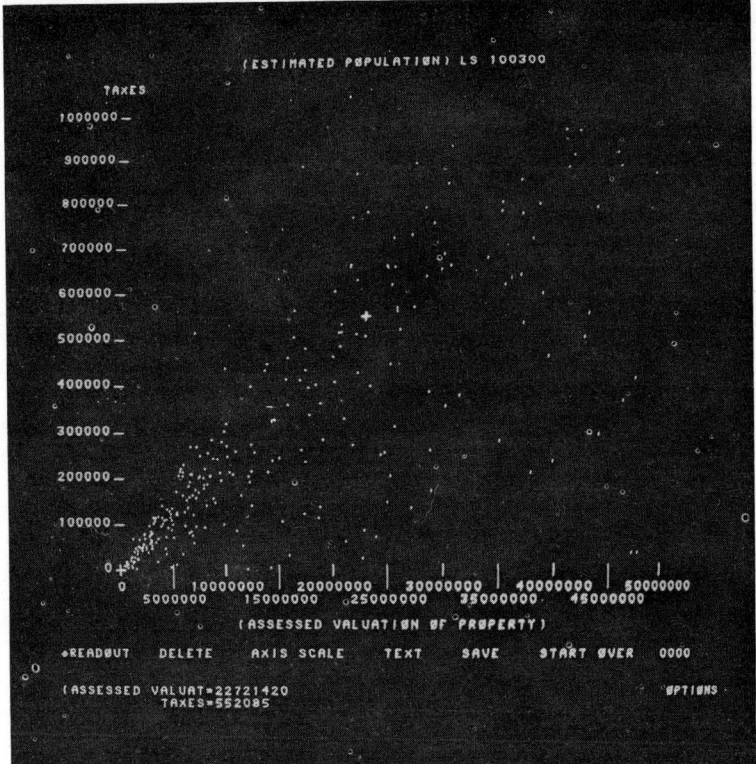

Fig. 3 Typical System Development Corporation data base display

Compression or tension in each member were displayed. With a light pen, the experimenter could modify the design by adding or removing structural members. The structure would be recalculated and the new values displayed at the appropriate locations.

Boeing has used computer graphics effectively to simulate the view through the windshield of an airplane to study the visibility problems during refuelling operations. On-line problem solving is a kind of simulation. The Bolt, Beranek, Newman Teleputer System which has been in operation for the last few years, allows conversation between the scientist (or student) and the data base, allowing them to solve problems by computer aided techniques.

Dr. Edgar Meyer, formerly of M.I.T. and now with Texas A. & M., and Professor Cyrus Leventhal, formerly of M.I.T. and now with Columbia, have developed programs which display crystallographically determined structures on a C.R.T. Inputs to the system include standard published experimentally determined parametric data (name and coordinates for each atom in the asymmetric unit, as well as cell constants and space group number from this International Table, together with connectivity tables). A molecule is represented by connecting bonded atoms with a visible line. Also, the asymmetric unit or the contents of the unit cell may be displayed together with the edges of the cell. The perception of the third dimension is greatly enhanced by slowly rotating the display. It is expected that this technique which is now experimental, will find significant application among the pharmaceutical companies. Great insight into chemical structures can be achieved with this kind of simulation.

Closely akin to the simulator applications are those in which the display is used to monitor simulated or on-line tests. For example, the graphic displays allow comparison between standard test data and the actual test data being received. One such system is installed in the NASA Space Flight Center and this computer controlled dynamic test system provided vital data on the Saturn IV Moon Rocket structural response to the stresses of flight. Systems Engineering Laboratories designed and built this system to simulate the vehicle flight dynamics in a laboratory environment. Data from shake testing was used to prepare mathematical models for simulation of actual flight conditions for both the vehicle and guidance systems.

IDI has installed similar graphic systems at ARO for on-line monitoring in wind tunnel tests.

Process Control

Two power companies in the United States, one in San Antonio and one in Houston, are in the process of installing C.R.T. displays to be used instead of the traditional status board common in the industry.

David E. Weisberg recently described typical applications of display consoles in process control.[8] Graphic equipment can be used in closed-loop processes (as typified by refining units in the oil industry, nuclear reactors, chemical plants, paper making machines, cement kilns and blast furnaces) to monitor past and projected future action of several variables. Figure 4 shows a representative display of this type. Open-loop processes, requiring operator intervention (such as a chemical or oil movement system) can be controlled by a graphic console using presentations like those shown in Figure 5. In such systems, the operator can control pumps and valves by pointing a light pen at the symbolic representation on the C.R.T.

The computer accepts this command, sends out appropriate signals to accomplish the operation and displays the results by modifying the movement pattern. Without the line drawing console, the status display would usually be wall mounted. In a very large system, it would probably be limited in detail with the status of valves, pumps and switches indicated by coloured lights. The flexibility of the C.R.T. display system permits the operator to use a large or smaller section of the total process, as he desires. The amount of detail increases as he works in smaller sections.

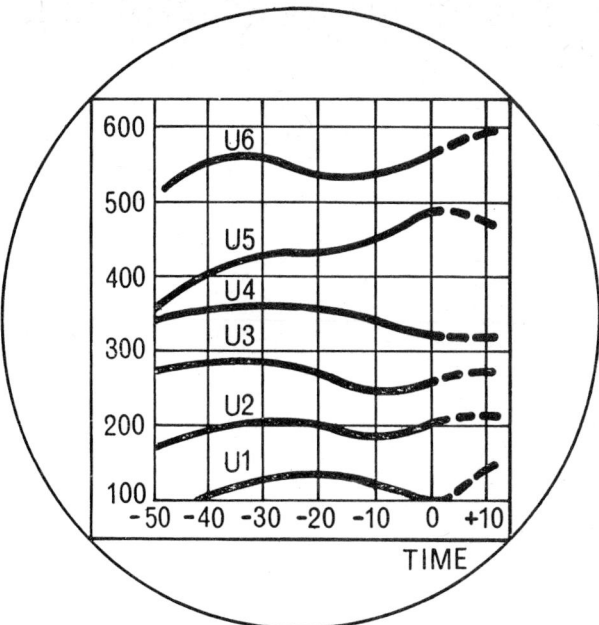

Fig. 4   Typical closed-loop system display (from Reference 8)

Computer Aided Education

In the last two years, there have been several programs in computer aided education under development, where the individual student interacts with the computer via a C.R.T. Now, there are at least three such public school experimental installations in the United States. At the Brentwood School in Palo Alto, California is an installation using the IBM 1500. The New York School system is engaged in a program with RCA and Philco-Ford has installed a system for the Philadelphia schools.

Two major problems which seem to stand in the way of broader acceptance of C.R.T. terminals in the computer aided education environment are (1) the writing of programs, (2) the terminal cost. Some estimates indicate that the terminal costs give rise to an expenditure of about two or three times that normally encountered in the usual elementary school environment. However, the costs become much more competitive when considered in the university environment and become very attractive when compared with the costs incurred in specialised training situations, such as the military.

Each program seems to be directed at somewhat different levels. For example, the program at Brentwood involves experiments with a group of slightly more than 100 first graders, while the Philco program is at the high school level and is currently used in teaching 10th grade biology and 9th grade remedial reading.[9]

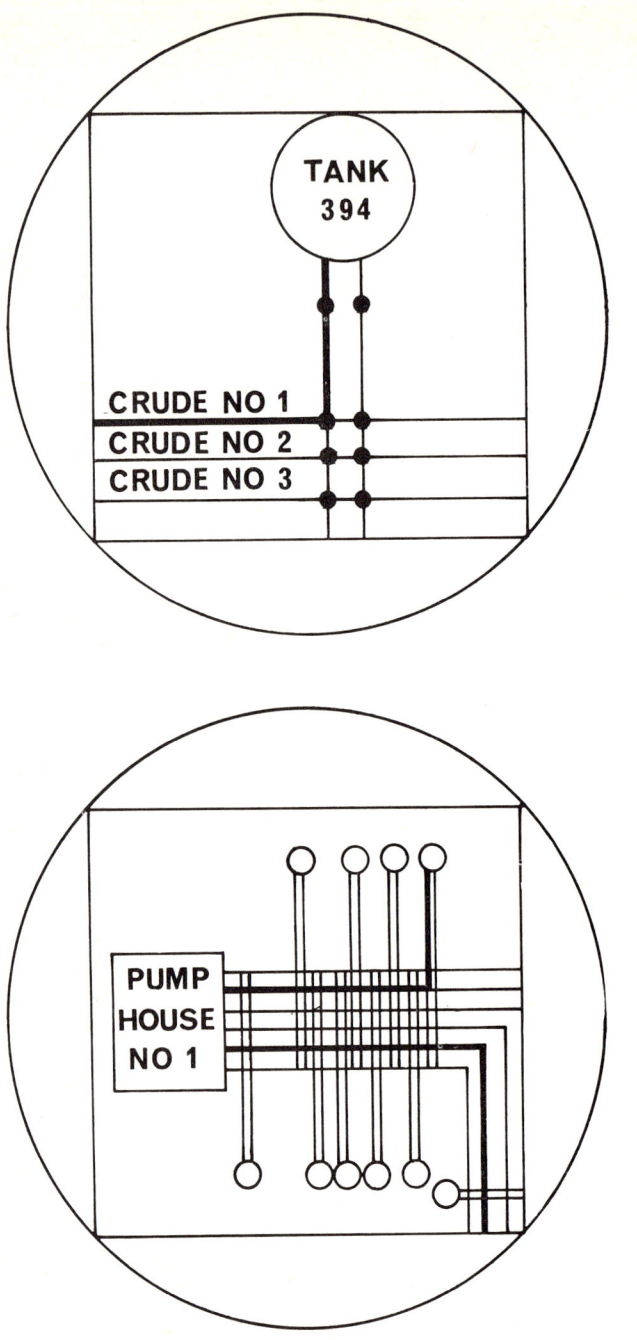

Fig. 5 Use of graphic display console in open-loop system (from Reference 8)

Several universities, including UCLA, Harvard, University of Wisconsin and University of Minnesota, have continuing programs of computer aided education.

Pattern Recognition

Computer graphics is helping to solve the general problem of pattern recognition. Various kinds of film (reconnaissance, X-ray, or bubble chamber, for example) are digitised and the data processed by computer. Using various correlation techniques, programs are developed which allow the computer to abstract significant information from the pictures. One may wish, for example, to detect man-made objects or special particle tracks, or the presence of a tumor. The display is useful because it permits rapid comparison between the raw data and the computer decisions. In more sophisticated systems, the programs can be refined on line by the experimenter using his light pen. Scientists at Albert Einstein College of Medicine and Columbia University are working with these techniques. A bit more than 1984ish, perhaps, is the use of a display to monitor the performance of experimental robots.[10] At Stanford Research Institute in Menlo Park, California, for example, a "baby" robot is learning to navigate in a cluttered room. Although, at the moment, the experimental machine can only move from one point to another through a room full of obstacles, it is hoped that in the future the device can be developed to carry out much more complex instructions, such as moving all the waste baskets from Room 217 to Room 321. It will be necessary for the robot to recognise objects and a display is being used now to monitor the performance of the robot in this respect.

Graphic Arts

From many indications, a revolution is quietly taking place in the graphic arts industry. The third generation of equipment involving cathode ray tube displays and film processing systems is predicted by some to mean the death of hot lead within the next five years. Characteristic of these systems is the ability to produce extremely good quality characters at high speeds. The RCA Videocomp, for example, can put 1000 characters per second on film or photosensitive paper. These rates mean that a complete newspaper page can be composed in about 10 minutes. Systems are also produced by Alphanumeric Inc., Harris Intertype in a joint effort with CBS Laboratories, and Mergenthaler. Business Week estimates these electronic devices have made photocomposition a $250,000,000 a year market.[11] In the future, some experts expect that graphic consoles will begin to play a part in the edit and compose functions.[12,13]

Art directors and illustrators will probably work on the face of the C.R.T. displays rather than on paper. Although final art work may still be prepared by these conventional materials and inserted into the system by electronic scanner, a C.R.T. display unit would be excellent for sketches and preliminary studies. The art work could be displayed simu'taneously in several newspaper editorial offices, discussed by telephone and revised on the spot by any of the participating editors. Further, the art director could insert graphic specifications into the system from his console for accurate page make-up and could receive proof for verification within seconds. He could either review these proofs on his display screen or obtain a hard copy print from a character generating proving unit.

Computer Generated Movies

Bell Laboratories is playing a leading role in the production of computer generated movies,[14] some of which can be characterised in the simulation field. Among the films that have been produced include films on harmonic phasors, produced by Professor William H. Huggin of Johns Hopkins University, and Professor Donald D. Weiner of Syracuse University. This film concerns the composition of complicated periodic wave forms by adding projections on an axis of rotating vectors or "phasors". The result is the presentation of a basic lesson in electrical engineering done in a way that is much more graphic than can be done on a blackboard. One of the classic computer films was first produced by Bell Laboratories in 1963 by E. E. Zajac and showed the result of the simulation of the motion of a communication satellite. Another film produced by D. E. McCumber was especially useful to the investigator in studying Gunn-effect instability, which had been implicit in, but not immediately apparent from, the mathematical description of the phenomenon.

The Boeing Company, Los Alamos Scientific Laboratory and Lawrence Radiation Laboratory have also been very active in the areas of computer graphics movies. A group has been formed in the United States which actively supervises the production of films in several universities. This group, The National Committee for Electrical Engineering Films (NCEEF) is chaired by Professor John Brainerd of the University of Pennsylvania.

One of the attractive features of computer generated films is the relative economy. The technique makes feasible the production of some kinds of films which previously would have been far too expensive and/or difficult. Typical films so far produced have fallen in the range of $200 to $2000 per minute. The cost of corresponding hand-animated films would have been at least twice as much in the simpler cases, and in other cases would not have been possible at all without a computer.

Other

In addition to the specific applications described earlier, it will be of interest, I think, to review some of the uses for graphic consoles being investigated by many universities in the United States. This list is representative rather than inclusive, and is meant only to give an indication of the thrust of academic research:

Albert Einstein College of Medicine . . . medical research.

Brown University . . . 3D presentation.

University of California (Berkeley) . . . graphic consoles are being used in the development of a natural language, as well as in a research environment.

UCLA . . . computer aided problem solving and instruction.

Columbia University . . . pattern recognition in conjunction with nuclear research.

Harvard University . . . computer aided problem solving and instruction.

University of Illinois . . . a chapter by Professor W. Gear in this book describes a project using a graphics terminal for program writing.

Illinois Institute of Technology . . . on-line problem solving.

Massachusetts Institute of Technology . . . general problems of man-machine interaction.

University of Michigan . . . development of programming languages.

New York University . . . computer aided design.

University of Ohio . . . nuclear research.

Pennsylvania State University . . . underwater technology studies.

Reed College . . . medical research

Stanford University . . . computer aided learning.

University of Utah . . . displays are being used in computer aided design, architectural design, and medical research.

And there are more applications of graphic consoles. For example, music is being composed; animated cartoons are being drawn; aircraft proposals are being developed; and proposed building designs are being shown to prospective clients. Implicit in these, and many other applications, is the use of graphic consoles in the time-sharing environment.

HARDWARE

C.R.T. Terminals

There are some 17 United States manufacturers producing direct view C.R.T. consoles suitable for computer graphics. See Table 1.

To this could be added a number of manufacturers producing alphanumeric systems . . . but this discussion will be limited to graphic consoles.

The performance range of these systems is indicated by Table 2.

TABLE 1

U.S. Manufacturers of Commercially Available C.R.T. Graphic Terminals

Adage.
Bolt, Beranek & Newman, Inc.
Bunker-Ramo Corporation.
Computer Displays, Inc.
Control Data Corporation.
Digital Equipment Corporation.
Information Displays, Inc.
Information International, Inc.
IBM.
International Telephone & Telegraph Corporation.
Philco-Ford Corporation.
Sanders Associates.
Scientific Data Systems, Inc.
Stromberg-Carlson Corporation.
Systems Engineering Laboratories, Inc.
Tasker Instruments Corporation.
UNIVAC.

TABLE 2[15]

Range of Performance
Commercially Available Graphic C.R.T. Terminals

Function	Quantity per frame (Based on 40 Frames/Sec.)
Random Dots	250–8300
Incremental Dots	2500–25,000 (assuming random, not raster, scan)
Random Characters	125–5000
Incremented Characters	220–8300
Connected Line Segments (Fixed Time Vector Generator)	160–830 (depending on Vector Generator, this can represent up to 16,000 inches per frame)
Connected Line Segments (Proportional Vector Generator)	160 inches – 50,000 inches
Circles	80–250
Price	$20,000 to $300,000

NOTE: Quantities shown exclude Display Generator, Computer and Memory Logic and/or cycle times, which can reduce data per frame by up to about 50%.

Ronald A. Siders presented an excellent summary of the requirements for an active graphic system.[7] He states, in part:

"The first requirement for an active graphic system is to accept graphical input on the scope face. A variety of capabilities of this type have been developed. Existing systems currently accept points, lines, circles, general conics and free-form lines directly on the scope. Geometrically perfect forms can be defined by inputting parameters. Scale can be changed by almost any useful factor. Elements of drawing can be deleted. The drawings can be rotated about any axis giving the illusion of three dimensions. Dimensions can be automatically calculated and displayed. Compounds can be duplicated as often as desired and placed in mathematically correct positions (such as teeth on a gear). Orthographic projections can be interchanged with perspectives and isometrics. One of the most powerful capabilities is that of calling up from core a wide range of standard components to add to the design on the scope. A circuit of a standard electronic component can be rapidly laid out. At any point, the design worked on can be filed away in memory for later recall. In this fashion, parts of the overall design can be worked on independently and called up for examination single or as a whole."

During the past two years, several distinct trends seem to be developing in commercial hardware available for computer graphics.

Integrated Systems

Several manufacturers produce integrated systems. These are fully buffered consoles which use a small digital computer, both as the display buffer and to do display "housekeeping", so that the load from the central computer can be reduced. Included in this category are the IDI IDIIOM, the IBM Model 2250 MOD 4, the SEL 816A, the Adage Graphic Terminal, Digital Equipment Corporation's 338 and 339 Programmed Buffered Display, the CDC Digigraphic series, and the Bunker-Ramo BR90. With the exception of the Adage Graphic Terminal, all of these units use digital computers as the programmable memory. The Adage Terminal uses a hybrid computer. Some systems such as the IDI IDIIOM, include a number of digitally controlled analog function generators, such as, character, line, and circle generators. Others, such as the CDC Digigraphic System, use point plotting displays that depend on software to generate graphic images. Because these units range in price from about $60,000 to about $200,000, they can be classed as medium to high cost systems.

Figure 6 shows one of these typical integrated systems and Figure 7 is a typical block diagram.

Lower Cost Terminals

Many workers, and potential users in the display field, have felt that broader usage of graphic consoles has been limited by their relatively high cost. In an effort to reduce terminal costs, several approaches have been explored.

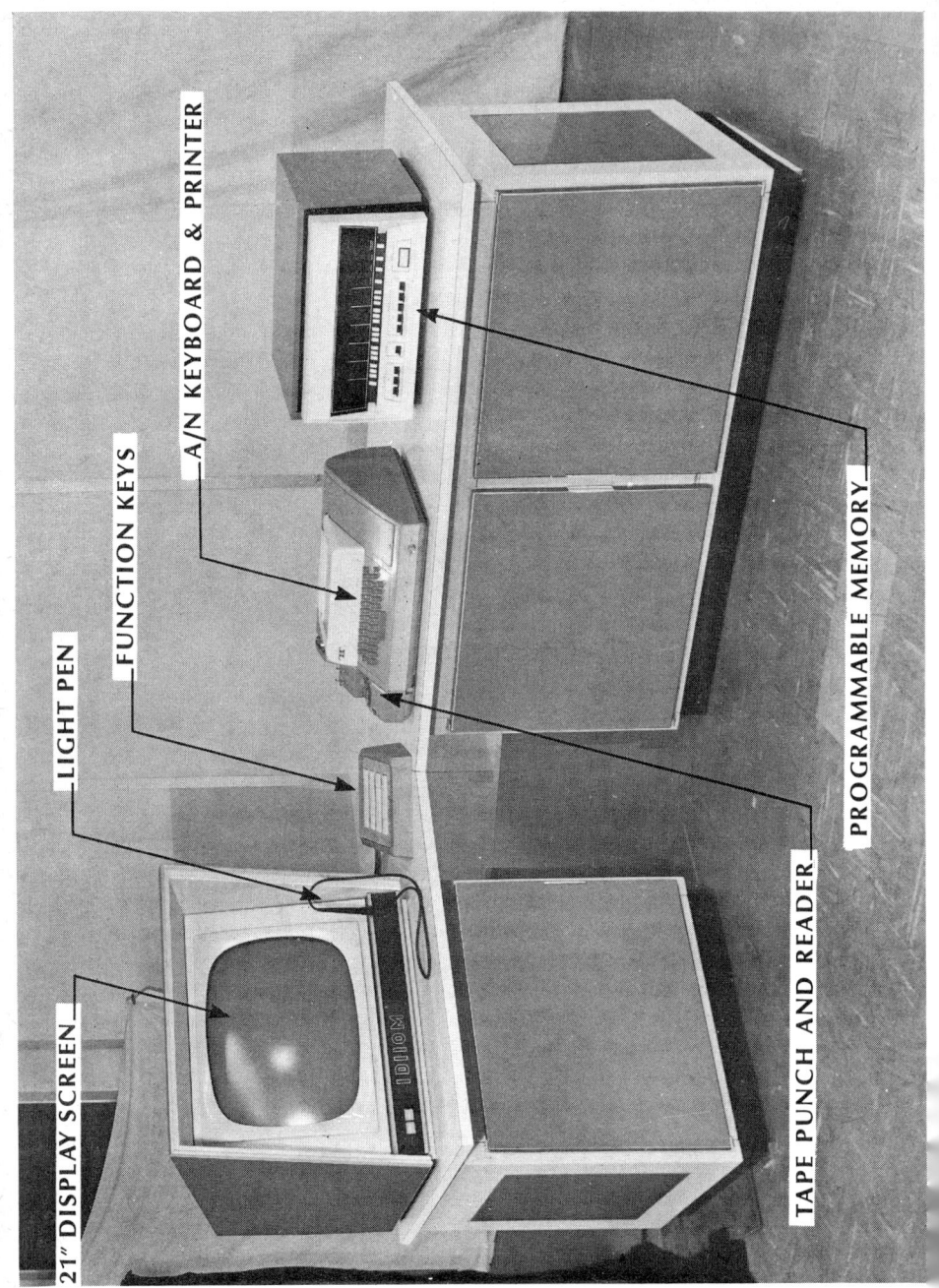

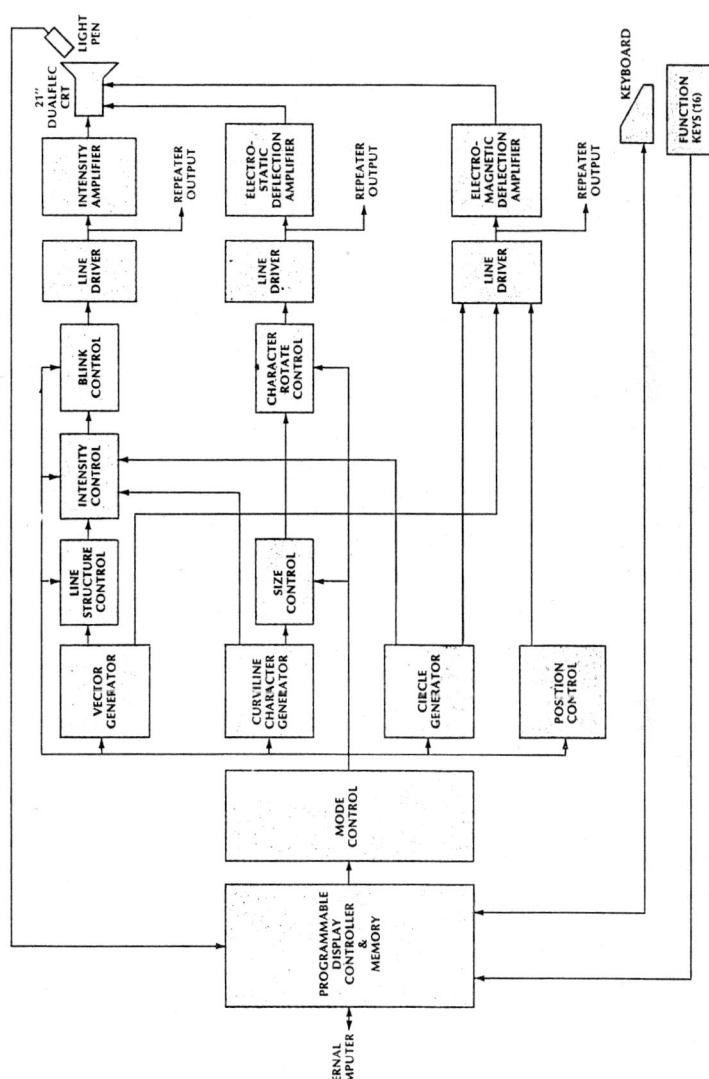

Fig. 7 Graphic console block diagram

Storage C.R.T.

When Tektronix introduced their $2000 Type 564 Oscilloscope with a 5 in. storage C.R.T., several groups began to use it as the heart of a low cost graphic terminal. One of the pioneers was the Bolt, Beranek & Newman Teleputer System (Figure 8). AMTRAM (automatic

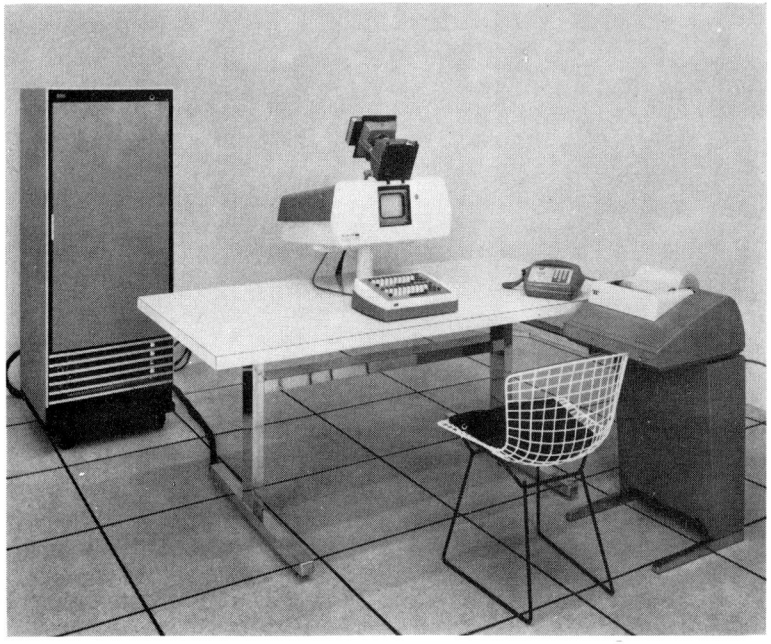

Fig. 8 Early system using storage C.R.T.—the Teleputer (by permission of Bolt, Beranek and Newman Inc.)

mathematical translation), a NASA development, was another interactive display system using the small storage C.R.T.[16] M.I.T.'s Electronic Systems Laboratory invested in the development of the low cost graphics terminal which would hopefully sell on the $5000 to $15,000 range.

However, although the terminals were low cost, the small C.R.T. tended to limit its use in graphic applications.

Within the past year, Tektronix has introduced a new larger screen (11 in. diagonal) storage scope, the Type 611.* This scope sells for approximately $2500. And a new company has been formed, Computer Displays, whose intention it is to produce commercially the M.I.T. design, using the larger Tektronix scope. In fact, also, the Digital Equipment Corporation has just announced the availability of the VD8/1 Storage Tube Display Controller. This relatively low cost controller provides the necessary conversion between the digital signals

* Editors' note: see Equipment Section, page 229.

from a small computer, such as the PDP-8, and the analog signals required to drive the storage oscilloscope, such as the Tektronix 611.

In spite of some limitations in the storage tube approach (difficulty of interaction and the presentation of rapidly changing data) the low cost should make it attractive for several applications.

татаТV Systems

The low cost of a standard TV Monitor has generated interest in producing a graphic display using the TV set as the output. Interactive graphic TV Systems have been produced by Hazeltine, Computer Control Inc., Monitor Systems, and Philco-Ford. However, to date, when the necessary programming is considered, costs have been high.

Low Cost Random Position Monitors

Low cost, wide bandwidth, large screen (17 in. diagonal) X–Y C.R.T. plotters have been offered by Hewlett Packard (HP 1300A), and Fairchild (736A, no longer in production). Experimental systems, using incremental stroke techniques and wide bandwidth high capacity discs (as the refresh memory) to drive these X–Y plotters, have been developed by Bell Telephone Laboratories, Friden and several research groups. None has yet been offered for sale, however.

Hard Copy Systems

Besides interactive, direct view graphic C.R.T. consoles, another widely used family of computer graphic devices is cathode ray tube plotters. About 100* are currently installed in the United States. Here, the speed of the C.R.T. allows rapid outputs from rapid translation of computer language to pictures . . . much faster than one can expect to get from mechanical plotters, although certainly not as accurately. A number of manufacturers now offer these on-line or off-line systems. Systems are offered by Stromberg-Carlson (the SC4020 and SC4060), the GEO Space Corporation (TP203), California Computer, Benson-Lehner, CDC (DD80), and IDI. These devices generally incorporate a 16 or 35 mm camera and some kind of immediate hard copy capability, using electrostatic or photographic paper (such as the 3M dry silver paper). Units range in price from approximately $50,000 to $300,000.

Color

Multicolor C.R.T. displays have been of interest. From a practical point of view, only one manufacturer currently offers to the industrial market a C.R.T. terminal with color capability. This is Philco-Ford's series of basic alphanumeric inquiry units that have some graphic capability. These displays use conventional color mask tubes. An

* This is another "educated" guess—C. Machover.

experimental graphic system (random positioning) was delivered to the Air Force by Digital Equipment Corporation several years ago. Hazeltine has produced a color TV system for military inter-traffic control applications.

Last year, Sylvania announced a color tube operating on a completely different principle. Instead of having three guns, each of which is directed toward an appropriate aperture, the tube depends on beam penetration for its color. At one depth of penetration, green is produced and at another depth, red is produced. This type of CRT is also available from Thomas Electronics. Much work still needs to be done to solve the high speed switching and dynamic focusing problems associated with this kind of tube. General Electric has just announced the avail-

How On-Line Graphic-Input Devices Compare

Device	Principal Advantages	Principal Disadvantages	Resolution	Input From Hard Copy	Cost*	Comments
Light Pen	"Naturalness" of use with CRT display	Tracking required; no hard-copy input	Low	No	Low	....
Analog Sheet Encoder	Hard-copy input; no tracking required	Analog drift; low accuracy	Low	Yes	Low	Some versions not commercially available
Digital Pressure-Sensing Encoder	Uses ordinary pencil or pen	Potential mechanical failure	High	Yes	Likely to be high	Not commercially available
Rand Tablet	Hard-copy input; no tracking required	Small size	High	Yes	High	....
Lincoln Wand	Three-dimensional	No hard-copy input	Medium	No	Likely to be medium	Not commercially available
Rho-Theta Transducer	Can accommodate large drawings	Requires polar-to-Cartesian transform	Medium	Yes	Low	....
Datacoder	Low cost	Mechanical encumbrance	Medium	Yes	Low	....
Track-Ball & Joy Stick	No tracking required	No built-in positional feedback; no hard-copy input	Medium†	No	Medium	....
SRI Mouse	Speed and naturalness	No hard-copy input	Medium†	Adaptable	Low	Not commercially available

*Not including display and/or A/D converters, if required.
†Basically cursor-positioning devices; as such, do not require high resolution.

Fig. 9 Summary of graphic input devices (from Reference 17)

ability of one of their aphanumeric terminals, the Datanet 760 with this Sylvania type color tube.

Input Devices

One of the powerful features of C.R.T. graphic consoles is the provision for operator input of basically pictorial material. This capability is in addition to alphanumeric or function key inputs.

A number of input devices have been developed and some are now commercially available.[17] Figure 9, reproduced from Reference (17) is an excellent summary of these devices. Most commonly used is the light pen. Becoming quite popular now is the graphic tablet, of which the RAND unit (produced commercially by the Data Equipment Division of Bolt, Beranek & Newman), is one example and the Data Tablet produced by Sylvania is another.

In addition to these well known units, several other interesting devices have been reported. Tasker Instrument Corporation has demonstrated a transparent pressure sensitive matrix over the face of the C.R.T. This allows the user to indicate a point on the face of the tube simply by pressing at the desired location. Early versions of the General Motors graphic system, DAC, used a transparent conductive coating on the face of the C.R.T. so that the location of a stylus would be determined by the stylus voltages.

Three-Dimensional Displays

Interest in three-dimensional displays seems to have diminished somewhat. Work is still being done in the generation of two slightly displaced images that can be viewed as a stereo pair. However, with the possible exception of the use of holography to develop three-dimensional images (principally by Conductron Inc.), there appears to be very little current U.S. work or interest in 3D.

Software

Commercial availability of hardware for use with the various applications described earlier continues at an accelerated pace. Special mathematics have been developed to facilitate the solution of some computer aided design problems.[18] However, software perhaps present a different and not necessarily encouraging picture. A number of different languages suitable for the generation and display of pictures, have been developed. Other languages have been developed for analysis or interpretation of pictures. What is needed, however, is a single language to handle these problems while containing features useful in a general purpose language.

An excellent summary of graphic language requirements is contained in a recent ACM article by H. E. Kulsrud.[19]

There is not as yet a common universally accepted graphic language and it would be foolhardy, I think, to make a prediction about either when such a language will be available, or in what form the language will be. It appears that the development of graphic languages has

G

followed the Tower of Babel that occurred early in computer languages. Each experimenter and user has individually developed ideas and each is not quite ready yet to merge these into a common, universally accepted language.

Summary and Forecast

In summary then, the state of the art of graphics in the United States may be characterised as in a transition state, between experiment and operational use. In such fields as aircraft design, integrated circuit design, and high speed printing, the profit making aspects are becoming more common. Systems which can be characterised as low cost graphics are just becoming available.

The thrust of future developments, it seems to me, has three directions; the use of color in graphic terminals, the further refinement of low cost terminal devices; and the development of a common graphic language.

Acknowledgements

I acknowledge, with grateful thanks, the contributions of technical material, and photos made by the following companies and individuals:

Sally Bowman—System Development Corporation
E. L. Campbell—RCA
Professor Steven A. Coons—MIT (Project MAC)
Frank Hagan—Digital Equipment Corporation
Professor B. Herzog—University of Michigan
John D. Joyce—General Motors Corporation
Dr. H. Kasnitz—MIT (Lincoln Laboratories)
Kenneth C. Knowlton—Bell Telephone Laboratories
Dr. Edgar Meyer—Texas A and M
George Micheals—Lawrence Radiation Laboratory,
    Boeing Airplane Company
C. B. Rogers, Jr.—IBM
R. E. Wye—Philco-Ford Corporation

REFERENCES

1. Sutherland, I. E. "Sketchpad: A man-machine graphical communication system", TR-296. MIT Lincoln Laboratory, Lexington, Mass., January 1963.
2. "Lockheed using computer-design", *Electronic News*, Monday, April 1, 1968, p. 36.
3. Jacks, E. L. "Observations on a graphic console system", General Motors Corporation (unpublished paper).

4. "Liberty, Motorola give buyers LSI lowdown", *Electronic Procurement*, September 1967, pp. 32–33.
5. "Computers at HemisFair '68", *Computerworld*, April 10, 1968, p. 8.
6. "Technology and man-power in design and drafting, 1965–75". United States Department of Agriculture publication.
7. Siders, Ronald A. "Computer-aided design", *IEEE Spectrum*, November 1967, pp. 84–92.
8. Weisberg, David E. "Man-Machine communication and process control", *Data Processing Magazine*, September 1967, pp. 18–24.
9. Charp, Sylvia, and Wye, Roger E. "An overview of project GROW—a computer assisted instruction system for the school district of Philadelphia" (to be published in *Educational Technology*).
10. " 'Baby' robot learns to navigate in a cluttered room", *New York Times*, April 10, 1968, p. 49.
11. "Printing is turning the page", *Business Week*, September 9, 1967.
12. Friedlander, Gordon D. "Automation comes to the printing and publishing industry", *IEEE Spectrum*, April 1968, pp. 48–62.
13. Tewlow, Jules S. "Time-sharing and the newspaper of tomorrow", ANPA Research Institute, Inc., Bulletin 951, April 15, 1968 (Part II).
14. Knowlton, Kenneth C. "Computer-animated movies", Bell Telephone Laboratories, Inc., January 1968. *Proceedings* of the Conference "Emerging Concepts in Computer Graphics", held at the University of Illinois, November 5–8, 1967.
15. Machover, C. "Graphic CRT terminals characteristics of commercially available equipment", *AFIPS Conference Proceedings*, Vol. 31, 1967 Fall Joint Computer Conference.
16. Wood, L. H., Ely, C. A., Glanzer, H., and Radice, V. "The AMTRAM input-output terminal", *Computer Design*, March 1968, pp. 68–74.
17. Keast, David N. "A survey of graphic input devices", *Machine Design*, August 3, 1967, pp. 114–120.
18. Coons, Steven A. "Surfaces for computer-aided design of space forms", MAC–TR–41 MIT Project MAC, June 1967.
19. Kulsrud, H. E. "A general purpose graphic language", *Communications of the ACM*, April 1968, pp. 247–254.

Biographical Note

C. Machover received the B.E.E. degree from Rennselaer Polytechnic Institute, and did his graduate study at New York University. During the past 17 years he has been concerned with the design and marketing of display devices, servo components, gyroscopes, bombing and navigation systems, and precision test equipment. He has been Manager of Applications Engineering for the Norden Division of United Aircraft Corporation and Sales Manager for Skiatron Electronics Television Corporation. He is now Vice-President (Marketing) of Information Displays Inc. and is President of the Society for Information Display.

The U.K. Scene*

F. E. TAYLOR
The National Computing Centre

The purpose of this chapter is to review the state of the display art in the U.K. First of all, one might well ask two obvious questions—how many displays will ultimately be used in the U.K., and where (and how) will they be used? The answer to these questions will determine future hardware product policy, and the type of software which will need to be implemented.

By the early 1970's there will be 2000 or so computers at work in the U.K. and each might well have several associated displays, depending upon the availability of time-sharing software. A reasonable estimate is, therefore, a "few thousand" displays, perhaps up to 20,000 in five or six years.

Within the computer field generally there is a clear interaction between appearance of hardware application and generation of software, events normally taking place in that order. Generation of software and improvements to hardware then extend application, which extends device markets and thus the funds available for further research and development. So far as displays are concerned, we are still effectively within the first cycle in the U.K.

FIELDS AND PROJECTED FIELDS OF APPLICATION

Alphanumeric Displays

Alphanumeric displays will be widely used for "softcopy" display of retrieved information such as that associated with:—

> *Hospitals*—patients' records, possible diagnoses, drug dosage. King's College Hospital are starting such a scheme.
> *Libraries*—for display of retrieved reference matter (both in public and industrial libraries). N.C.C. have two displays on order for this purpose—in particular for display of retrieved program abstracts in connection with the National Computer Program Index.
> *Management Information*—such as stock, sales and production control data in industry. One mail order house is already experimenting with a small number of displays for record handling and a large supermarket chain has just taken delivery of a system involving some tens of displays for stock control.

Such displays are useful in all cases where handcopy is not needed, or selection is required before hardcopy is produced.

* © N.C.C.

Further examples are:—

Airline Operation—for transmission of information such as available seats on an aircraft, passenger flight lists, cargo schedules, and loading information to maintain correct centre of gravity. The leading example in the U.K. is the BOADICEA system, which will ultimately use approximately 700 Ferranti displays. A small number of these are already being brought into use.

Query Systems—systems where a display is used in exactly the same manner as in the case of libraries, except that the information retrieved and checked may be operational, rather than reference information. Examples are: (i) a system projected by the West Midlands Gas Board, which will ultimately use something more than a hundred displays for checking of customer accounts on request. These are linked via a data transmission system, this being the key to the application of several displays at remote locations linked to a central processor—in this case via modern interfaces.

(ii) Plans by the East Midlands Gas Board for a similar system, on a similar scale.

Air Traffic Control—the LINESMAN/MEDIATOR Project incorporates displays for this purpose.

Experimental Results—a large aero engine manufacturer has acquired a number of alphanumeric displays for display of engine parameters during test.

Conversational Computing—the alphanumeric display may be used as an alternative to the teletype in a conversational computing system. The National Engineering Laboratory, East Kilbride, have acquired two displays for experimental use in this capacity.

Graphic Displays

Graphic displays with ability to display pictures as well as text will be applied where a diagram is conventionally used, or can replace a great deal of alphanumeric output—e.g. assembly drawings and other drawings required by engineering operatives. A second typical application of such terminals is in the field of process and system control—graphical data such as power flow along given transmission lines can be directly presented to the shift operator as a picture, rather than as text or a table, which take a greater time to assimilate. An example of such use of graphic displays is the C.E.G.B.'s new National Control System which will incorporate several Ferranti Display Control modules centred around three Argus computers. A smaller, similar system for the Midland area will incorporate Marconi units. Systems for the output of process control information and engineering drawings both fall within this category. The former are further exemplified by the displays associated with Dungeness "B" Nuclear Power Station (of the order of ten to twenty ultimately) and those which the British Steel Corporation plan to use in connection with automation of their cold steel rolling mills. (The Corporation also has systems including touch wire displays for steelmaking process control.) Such displays will not become widely used until design techniques using fully interactive graphic displays are established, thus creating the computer required for output to actual operatives on the shop floor. Work by members of the IBM 2250 Users' Group using a 2250 display to output the tool path when machining an engineering part is encouraging, as is also work described elsewhere in this volume by F. M. Larkin of the U.K.A.E.A. Culham Laboratory, using a display to show the paths of particles in a gas plasma when constrained by a magnetic field.

Interactive Displays

The third type of display is the interactive display, where interaction between operator and computer is possible via an interactive device such as a light pen, rolling ball, touch wire, or, if time permits, a keyboard. Computer response to operator action must be on a timescale acceptable to a human operator—within a few seconds at the most.

Such displays are already being looked into as a tool for design in several fields, for example:

General mechanical design. The National Engineering Laboratory, East Kilbride, currently have two, soon to be three, interactive displays for experimental work in this area.
Structural steelwork. e.g. for building frames and car bodies. One automobile manufacturer is currently experimenting with an interactive display for this purpose.
Aerofoil sections. Lockheed pioneered work in this area in the U.S., and Dr. J. V. Oldfield has recently demonstrated conformal mapping of a circle to an aerofoil using the Edinburgh University C.A.D. System.
Turbine blades, shafts, bearings. Work in this area by Rolls-Royce is described in another paper.
Mechanical linkages and cams. The APACE Centre at Aldermaston is active in this area.
Architecture. Work by West Sussex County Council is described elsewhere in this volume, and similar work was undertaken a year or so ago by W. Newman at Imperial College, which allowed an architect to view and interactively modify the layout of a proposed building constructed from industrial building modules.
Electronics. Selected circuit parameters can be input using a light pen or other device, and results returned to the same screen. Cambridge University and several IBM 2250 users are active in this area, in particular S.T.C. whose work is described in another paper. Racal Electronics have recently taken delivery of a display for circuit design applications.

Nuclear Physics. Work is being carried out in this area at Daresbury Nuclear Physics Laboratories, and Oxford and Birmingham Universities.

Chemical Engineering. Pipework and pressure vessel design. Messrs. Humphreys and Glasgow have announced a pipework system using a plotter which could be translated to use a display. A major chemical engineering firm is using a display for automated pipeline layout.

At present some three dozen interactive displays are in use or on order in the U.K. This includes research displays such as that being used in the Psychology Department of Sussex University and those being used for interactive mathematical surface generation and fitting at Cambridge University and Imperial College.

THE PRESENT "STATE-OF-THE-ART" OF U.K. SOFTWARE

Applications packages should ideally be machine independent, graphics software being linked to such programs via subroutine "call" statements.

I.C.L. are providing a suite of programs for their 4280 display, comprised of the DISMAN system providing DISplay MANipulation routines linked to ALGOL via a package known as EDGAR, and to FORTRAN via a package known as FRED. Data structure packages are also available.

I.C.L. are also linking their 1830 software to FORTRAN via a series of "call" statements. Each system supplies means for input/output, and operation of a light pen.

Marconi's have a different approach to the problem of generating display software. In brief, they have devoted a great deal of effort to generating a versatile ring data structure, such that tagged files within the data structure may be displayed by appropriate calls as either a main picture or a sub-picture. Input of information is currently from a specially extended standard keyboard, but software for linkage to other programs is anticipated later.

Imperial College have produced a "Network Definition Language",* aimed at avoiding the tedium of programming in a low level language, and the restricting framework of a high level language. Each program is regarded as a network and generalised servicing routines are provided as far as possible. A "Network Compiler" and "Reaction Handler" have been in use for some months.

In conclusion, the case for and against hardware and software standardisation might usefully be considered.

On the topic of standardisation, it is of interest to note that attempts made during World War II to standardise radar displays, were destroyed by manufacturers undercutting standards to gain a cost advantage. For such standards to take effect, some economic pressure is necessary. Standardisation can retard progress if it is performed ruthlessly at

* Computer Technology Group Report 67/7, Imperial College.

the expense of flexibility—for this reason a time limit of roughly three years has been set on the usefulness of any static standards which might emerge from an N.C.C. Working Party recently set up. Such standards may well be extended with time to take new developments into consideration. Alternatively, at the end of a given period a completely new specification for standards may be needed.

Against standardisation is the point that standardisation can push up the price of displays due to users having to pay for unwanted facilities—this is the complete opposite of the desired objective—a reduction in hardware cost and establishment of software which facilitates program interchange.

However, this field is still in its infancy, and a breakthrough in hardware costs could revolutionise the whole field.

Biographical Note

F. E. Taylor, Ph.D., leads those computing activities of the N.C.C. connected with interactive computing techniques, particularly those involving graphics and design automation.

Low Cost Graphics

MURRAY A. RUBEN
Digital Equipment Corporation

Most computer graphics systems in existence today are based upon the direct view, refreshed C.R.T. These systems, for the most part, are fairly expensive, generally costing well over $20,000. The recent availability of high resolution, moderate size, direct view storage C.R.T.s makes possible full computer graphics systems costing under $20,000. This chapter will compare several possible graphics systems based upon the direct view storage tube, and also some systems using the refreshed C.R.T. Emphasis will be on hardware and software tradeoffs permitting lowest system cost, and on the resulting performance and applications for such systems.

The basic properties of a random scan refreshed C.R.T., a sequential scan refreshed C.R.T., and a direct view storage C.R.T. display device are compared in Table 1.

Both refresh systems require a local memory to refresh the display. The major difference between the two systems is the manner in which the C.R.T. is scanned. The random scan draws the picture by tracing the lines on the display on a one to one basis. The sequential scan must first transform the picture into a T.V. raster scan.

The raster scan display is the least expensive output device since it can use the full benefits of television technology. Its use for computer graphics is limited almost entirely to alphanumeric readout and very simple graphs, because of the very difficult, time consuming, and expensive job of performing the space-time transformation of a general graphic picture into line sequential format, and because of the very large memory required for such a display. If, however, the application required many output displays (usually well over 16 scopes), all containing pictures processed by the same central controller, and using a common central bulk refresh memory, this approach can offer the least expensive total system. Typical memories used for the line sequential display include delay lines, magnetic core, disc or drum, video tape recorders and electrostatic storage tube scan converters.

The sequential scan system furthermore, is only practical when all the terminals are located near the central controller, since this system requires very high video bandwidths and hence cannot tolerate long transmission paths. Alternatively, Community Antenna Television (CATV) techniques can be used.

The quality of the sequential scan is probably the lowest of the three devices described in this paper. Linearity is often no better than 15%, and the resolution and information density may be as low as a few

TABLE 1

Some Properties of Random Scan Refresh, Sequential Scan Refresh, and Direct View Storage Tube Low Cost Graphics Systems

Parameter	Random Scan Refresh	Sequential Scan Refresh	Direct View Storage
Refresh Memory	high speed circulating and/or random access	high speed, circulating	built-in storage surface
Driver Circuitry	very high speed, high accuracy	high speed, low accuracy	low speed, high accuracy
Signal Bandwidth	high	very high	low
Source Memory Requirements	large	very large	small
Information Density	moderate	moderate	very high
Resolution	very good	poor	good
Brightness	good	good	low
Size	up to 18 in. sq.	up to 18 in. sq.	$6\frac{1}{4} \times 8\frac{1}{2}$ in.
Interactive Devices	almost any existing technique	generated cursor or light pen	generated cursor perhaps light pen
Response	immediate	intermediate, depends on processor speed	several seconds

hundred characters. The display itself will be bright, virtually any reasonable size, and flicker free since the frame rate is fixed.

The random scan refresh C.R.T. display is the form used most commonly on the more expensive computer graphics systems. Low cost systems using this technique will have to sacrifice information density to achieve the cost savings in less memory and slower drive circuits. Again, such a system only appears competitive when multiple terminals share the central memory. With random scan systems, the advantages of brightness and linearity are offset by the disadvantages of flicker but a higher information density than the sequential scan is possible with the use of hardware vector generators. The system controller is less expensive than the sequential scan system, thus permitting a system to require less terminals to achieve the same per terminal cost. The random scan system also offers the easiest means of graphic input from the C.R.T. via the light pen, a task much more difficult to achieve on the T.V. scan system. A second advantage of this random scan system is its ability to convey real time motion to its display. This is not possible in the T.V. system or the storage tube system because of the time required to re-format the picture.

The cost of the refresh system is governed basically by the complexity of its controller and the size and form of its refresh memory. Discs, delay lines, and core form the refresh memory. Highly interactive graphic communications, requiring either rapid response or very flexible graphic input, form the most likely applications for the random scan refresh system.

Both refresh systems discussed so far do not, at present, offer any immediate possibility of being capable of reaching a per terminal cost level for full graphics much below $10,000 per terminal in small terminal systems. The direct view storage tube (D.V.S.T.) systems to be discussed below offer the possibility of complete system per terminal costs below $10,000 now, and well below $5,000 within a few years.

As its name implies, the direct view storage tube has the property of retaining a visual image as it is written in the exact form in which it is written. This eliminates the requirements for a separate refresh memory and also greatly reduces the information rate requirements on the data source and its control and drive circuitry, since the image need only be written once. These two basic facts, when compounded, are responsible for the inherent low cost of the D.V.S.T. display system.

Figure 1 presents a block diagram of one form of computer graphics system built around the D.V.S.T. In this system, a small general purpose computer operates on line to one or more local display terminals. All control functions originate in the computer under command from the terminal keyboards. The D.V.S.T.'s are driven from an analog function generator controlled by the computer. Use of analog signals is the key to the low system cost, since they operate under conditions of high accuracy but low slew rate. These conditions are easily met by modern integrated circuit techniques. The second key

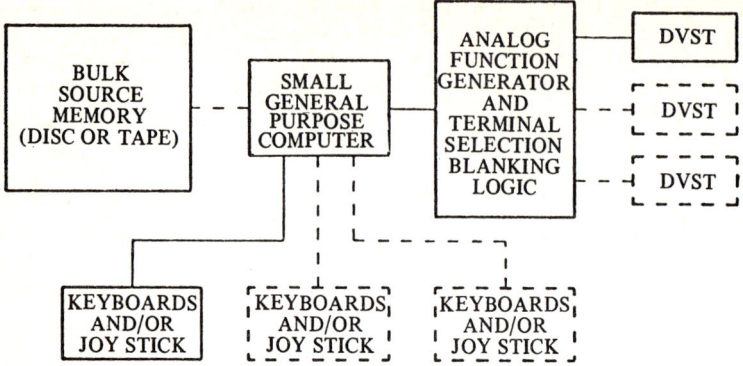

Fig. 1 Stand alone single or multi-terminal small graphics system using general purpose computer processor, a central analog function generator, and selectively blanked terminals. Scope signals are distributed at low speed up to several hundred feet in analog form by multi-conductor facility. Additional source memory in the form of tape or disc may be added for bulk information retrieval applications.

to low system cost is the heavy use of software, as opposed to hardware, to do nearly all data formatting and data control functions. Again, it is the relatively slow speed requirements of the D.V.S.T. which allows the use of software in this manner.

The system of Figure 1 can operate as a complete stand alone graphics system with the full power of a general purpose processor, for under $20,000. With a 500K word disc and 3 or more extra terminals, the total system averages less than $12,000 per terminal. When a 32K word disc is used instead of a 500K disc, costs drop below $10,000 per terminal.

The system of Figure 1 is highly attractive for general information

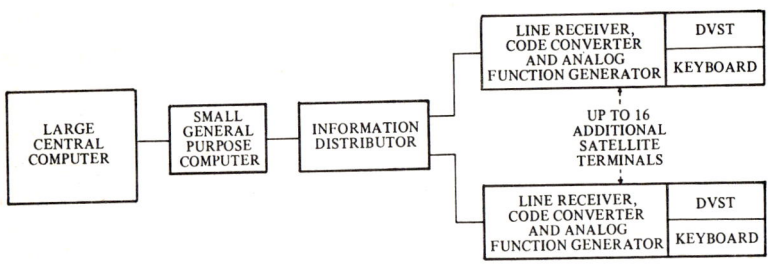

Fig. 2 Satellite processor graphics system using small general purpose computer as the central I/O controller and decoder for up to 16 satellite scope terminals. Transmission is by digital techniques over a high speed two wire facility between the scopes and the processor. Analog conversion is performed at the terminal proper. Communication between the central computer and satellite processor can be by direct link or by telephone facilities. Information rate in this link can be reduced by special coding and dual processor techniques.

retrieval and interactive graphic design in an environment permitting the local grouping of one or more terminals in the same room or adjacent rooms to the computer, and in an application where fairly rapid computer response in a shared environment is desirable.

Figure 2 illustrates a larger system version of Figure 1 in which the general purpose computer is now totally dedicated to controlling up to 16 terminals linked to it by coaxial or twisted pair lines. This system would operate into a large time sharing C.P.U. If wide bandwidth lines are used, the response speed is nearly as fast as the system of Figure 1, but per terminal costs would average around $8,000 not counting the transmission link and interfaces to the C.P.U., and not counting the C.P.U. and its associated peripheral costs. The system of Figure 2 is best suited for applications where all the terminals are within the same or adjacent buildings to the central computer. A system such as Figure 2 would be ideal for mass multi-port information retrieval or large computer aided design problems.

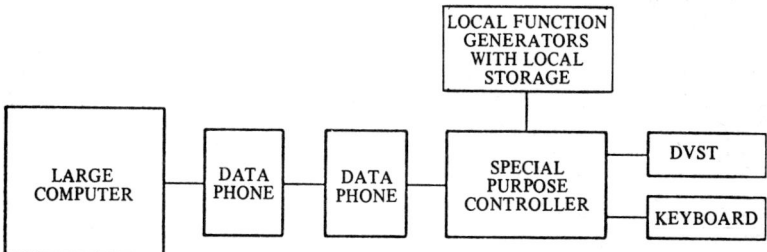

Fig. 3 Remote stand alone terminal system uses special purpose controller and local hardware function generators to overcome low bandwidth of transmission path. Communication is usually by ASCII or similar universal code over standard dataphone link to the central computer.

Figure 3 illustrates another approach, still using a D.V.S.T. The system is structured to operate directly on a telephone line. The information rate of this system, approximately one to two orders of magnitude slower than the direct system of Figure 1, represents the penalty paid for the use of low bandwidth lines. This reduced rate makes a hardware character generator necessary at the terminal, as well as other hardware required to format vectors and control the telephone interface and keyboard. All this special purpose hardware negates much of the favorable software/hardware tradeoff possible with the D.V.S.T., and produces a terminal costing between $10,000 and $15,000. To this cost must be added the dataphone and line charges, and the C.P.U. computer charges. The system of Figure 3 represents a solution to the remote access graphic terminal problem where it is not feasible to have direct lines to the computer. Also, with increasing use and advances in M.O.S. (metal-oxide-semiconductor) technology, the cost of a terminal as shown in Figure 3 can be expected to drop to the $7,500 range within a few years.

All of the D.V.S.T. systems have similar performance characteristics except for response speed. The D.V.S.T. now available produces a high resolution image (over 4000 characters) of a size, quality, and information density comparable to that found on a standard $8\frac{1}{2} \times 11$-in. page of text with 1 to $1\frac{1}{2}$-in. borders. The display itself does not flicker, but on the other hand contrast and brightness, while adequate, are less than that achievable with refresh systems. No existing refresh system can match the information density of the D.V.S.T., and it is in this area that truly the D.V.S.T. ranks supreme.

There are some applications where the use of a D.V.S.T. does present difficulties. Its use for true real time applications, requiring essentially instantaneous response, is not possible. Also, the present D.V.S.T. lacks selective erase capabilities, so that the entire frame must be re-written if it is desired to alter any substantial part of the picture. Here the re-write time of one to five seconds for a system like Figure 1 may not be objectionable, but the 30 to 50 seconds for the remote system might be. There are some program tricks which can be used to minimise the number of times the picture is re-written, such as the use of a special text buffer area when doing text editing, and the use of a partial refresh technique where the last few actions entered in the picture are drawn in a non-stored manner and are only stored upon verification by the operator.

This latter mode is possible in a bistable storage tube, where it is possible to observe a visual image written on the screen at a writing rate too fast to be stored by the tube. This image does not affect any previously stored images. Thus, the tube used in this manner may permit the use of a light pen or joystick to display and enter graphic data directly from the tube surface. The program to interpret this data, however, under some conditions is fairly complex. A technique which can be used in some programs is the tagging of what may be called target circles, or hot spots. The display is drawn with certain co-ordinates identified by a distinctive circle or other symbol. When a joystick is positioned so its generated non-stored spot appears at a target circle and the appropriate interrupt button is pressed, the program would read and identify the action associated with the target circle. Similarly, graphic input could also be read in from the joystick.

Before concluding this chapter, it may be appropriate to gaze into the crystal ball for future developments which would make possible even lower cost full graphics than indicated herein. There are two major technological advances which appear on the near horizons which can, together, result in price reductions down to the $3,000 to $5,000 per terminal total system costs. These are the rapidly emerging development of large scale M.O.S. integrated circuit arrays, for both logic and memory applications, and the development of photochromic glass direct view storage tubes. The former will make possible low cost memories and logic for character generators, vector generators and other logic associated with D.V.S.T. systems. Their relatively slower response compared to monolithic integrated circuits does not affect their use in D.V.S.T. systems, since these operate at completely compatible speeds. The photochromic storage tube is an extremely simple design in which chemicals in or on the glass face of an ordinary cathode-ray tube change their optical density when exposed to a beam of electrons. RCA in the States and Ferranti in the U.K. are working on these tubes.[1,2] Since no storage grid or flood electrons are required to maintain an image, the cost of storage tubes using this principle will be between one third and one fifth the cost of a present electrostatic tube. This can amount to $1,500 and more savings per terminal.

To summarise, this chapter has presented several forms of computer graphic display systems capable of per terminal costs below $20,000. Generally, refresh C.R.T. systems are more expensive and contain lower information density than do systems based on the direct view storage tube. These latter systems offer per terminal costs today below $10,000, but are only suitable for applications where system response times from a few seconds up to a minute are acceptable. However, such display systems can be expected to reach a price level below $5,000 per terminal within a few years.

Bibliography
(1) Phillips, W. Private communication (RCA).
(2) Turner, G. C. Private communication (Ferranti).

Biographical Note
Murray A. Ruben received the BEE degree from CCNY in 1962 and the Electrical Engineer degree from MIT in 1964. He is presently employed by Digital Equipment Corporation as a product development engineer.

Remote Display Terminals

R. ELLIOT GREEN
Scientific Control Systems Ltd

Abstract

The various types of display terminals used for remote on-line interaction with computer systems are investigated and details are given of their operation and applications. The methods of data transmission between the computer and the terminal are discussed and present research and development projects, both in this country and in the United States, are reviewed.

Introduction

The emphasis in the preceding chapters has been on interactive graphic terminals, which maintain their display by continually reading and actioning a set of commands held in the computer's core store. The data channel between the computer and the terminal maintains a very intimate link between the two, allowing a high rate of information transfer simultaneously in both directions. This data channel must be short and it would be uneconomic to design a terminal of this type to be operated at any appreciable distance from the computer.

Multi-access and time-shared computer systems, which can be entered and operated remotely are gaining favour very rapidly both here and in the United States. Several companies in both countries are selling computer time to clients, who merely hire a teletype terminal and use it over a standard phone line.

If an incremental plotter is used in conjunction with a teletype, the combination is a fairly versatile but slow display terminal capable of receiving both alphanumeric and graphic data.

The limitations on the speed of information reception are set by the mechanics of the devices themselves. This is no longer the case if one uses a C.R.T. terminal; here one is limited by the bandwidth of the phone line. The normal switched voice grade lines will in general transmit up to 1200 serial data bits per second (baud). This allows an alphanumeric C.R.T. terminal to display a message from the computer over 10 times faster than a teletype. Over a dozen manufacturers in the United States are selling display terminals of this type. They are character only displays, their graphic repertoire being confined to very rudimentary diagrams, which are built up out of characters.

Much research and development is going into attempting to bridge the gap between the expensive interactive graphic terminal, which cannot be separated from its driving computer, and these comparatively inexpensive but limited remote units. The first products of this new generation of remote graphic display terminals are now just beginning to appear on the market.

Data Transmission Techniques

A large multi-access or time-shared computer will require a local small satellite computer to act as a communications controller. This is controlled by the supervisor program in the main machine and in turn is programmed to control input to and output from the terminals. It interrupts the main computer to feed in data and to run user programs.

Both the computer and the terminals generate and recognise parallel binary data. For transmission across a single low bandwidth connection (two wires), it must first be changed from parallel to serial form and later decoded at its destination. This is done by a data adaptor unit at the computer end of the line and within the terminal at the far end. For transmission along the line itself a modem (modulator/demodulator) unit is also required at each end. These act as interfaces to the line changing the binary data into frequency modulated form suitable for transmission.

The G.P.O. offer data transmission services at various line speed ratings. For display terminals this should be as high as possible. The Datel 600 service operates over the public switched telephone network.

This offers a transmission rate of up to 1200 baud and is operated in half duplex mode, i.e. 1200 baud transmission in one direction at a time only, with a simultaneous supervisory channel (along the same two wires) for an interrupt signal at 75 baud in the other direction.

To use this service all one requires is to hire normal telephone connections and modems from the G.P.O. at the computer and terminal locations. Alphanumeric data is normally transmitted in asynchronous mode using standard ASCII (American Standard Code for Information Interchange). Seven bits are used to define the character and one as a parity check. Each character is preceded and followed by start and stop bits so the terminal does not need to be synchronised to the communications controller. If one allows for these and other overheads such as control characters, about 100 characters per second can be transmitted across a 1200 baud line. The data could also be transmitted in synchronous mode; this is more efficient than asynchronous mode as it does away with start and stop bits with each character. It however requires more sophisticated terminal equipment.

In both modes insertion of parity bit and checking for parity error is done in the data adaptor and by the terminal itself. There are two basic methods of parity checking; by character and by block of data. Block checking is the more efficient and normally includes a character check, but it requires more complex equipment.

The .G.P.O. is about to introduce a Datel 2400 service, which will operate in full duplex mode over hard wired dedicated lines. These are laid to order and their lease is fairly expensive (£500 p.a. for 20 miles).

The full duplex mode uses four wires and offers transmission in both directions simultaneously at 2400 baud.

If the terminal is situated "in house", i.e on the same site as the computer and within a mile or so, it may be more economical to connect it directly with a coaxial cable. This will allow it to be operated in a manner as described but with a data transmission rate an order better than with a phone line. (N.B.—In-house alphanumeric C.R.T. terminals are referred to as "local" terminals by several manufacturers.)

Techniques for transmitting graphic information are far less advanced. A standard incremental plotter describes pictures in terms of small pen movements. Each increment is defined by a code of 5 or 6 serial bits. As even a simple diagram could require some ten thousand increments, direct application of this technique to C.R.T. terminals would necessitate very slow drawing rates and large expensive buffer stores. Thus most operational remote graphic C.R.T. displays to date have been connected to the computer by coaxial cables.

Teletype and Incremental Plotter

The teletype is now a fairly common and relatively cheap (about £500) device and requires little description. It is a teletypewriter, which, when connected to the computer, can send a character on key depression, and can receive and type at about 10 characters per second. Editing facilities on present systems are usually limited to deletion of the previously typed character or characters.

There are two basic forms of incremental plotter: drum and flatbed. Drum plotters draw on continuous edge perforated stationery up to 30 in. wide, and their prices range from £2,000. A pen is moved across the paper, which is moved longitudinally from one roller to another. All moves are increments either by the pen or by the paper or both together, and are under computer control. Increments are usually of the order of $0 \cdot 005$ in. and are effected at a rate of about 300 per second, with positional accuracy of about $\pm 0 \cdot 005$ in. Either ballpoint or liquid ink pens can be used giving a wide variety of line widths and colours.

Flatbed plotters differ from drum plotters in that the pen is moved in both x and y directions over a stationary sheet of flat paper. In general they are more expensive (£8,000 upwards) and offer greater precision ($\pm 0 \cdot 002$ in.). They will allow the use of non-standard stationery—e.g. geographical maps, on to which are superimposed weather maps. Some of the more sophisticated flatbed plotters offer a method of inputting positional data to the computer.

The computer can be remotely interrogated by use of the teletype. A program could be typed in, fed in on paper tape or called up from the computer mass store. Data could then be fed in via the keyboard and the computer's reply appear either in alphanumeric form on the teletype or in graphic form on the plotter. The two devices are under the control of a single control unit, which is on-line to the computer.

Applications of Teletype and Incremental Plotter as a Remote Graphic Terminal

The teletype alone has many applications as a remote alphanumeric i/o device. Programs can be called up from mass store, fed in on punched paper tape, written on line and run in an interactive mode. Their use by customers of computer time sharing companies is increasing rapidly, and applications range from calculation of discounted cash flows to solution of complex differential equations.

The combination of plotter and teletype is of value where any quantity of numerical data is required from the computer. It allows a dynamic solution to a differential equation (e.g. a pressure against time graph) to be plotted. Data can be output far more quickly in this form than a numerical printout. It is also reduced to a far more comprehensible form. In fact, by transforming numerical experimental data into automatically plotted graphical form, the terminal can save the scientist or engineer hours of unnecessary work and can add a new dimension to the power of his experimental techniques.

This type of remote terminal is being used widely in the United States in many fields including:

> financial analysis
> advertising and market research
> test data graphing
> critical path network scheduling
> atomic structure analysis
> medical and psychiatric diagnosis and research.

Alphanumeric C.R.T. Terminals

These terminals are, in effect, somewhat more sophisticated alternatives to the teletype. Their prices range from £2,500 to about £10,000 depending upon the sophistication required and the system configuration. Their value lies where rapid and convenient input and/or output is required to on-line computer systems. They basically comprise an input keyboard and a cathode ray tube display screen with control logic and buffer data store. Messages typed on the keyboard appear on the screen, as do replies from the computer. Most terminals use a modified form of raster scan with a monoscope, dot or starburst matrix. Some manufacturers use a standard TV set with a scan converter.

Facilities Offered by Alphanumeric C.R.T. Terminals

Input of data and commands to the computer are via a keyboard similar to that of an electric typewriter. In general, this contains about 64 character key functions, which include upper case letters, numbers, punctuation and special symbols. There are usually also about twenty control keys, which control commands to the computer, editing functions and cursor positioning. This cursor is a small marker, which is displayed on the screen in the position where the next character is to be typed. On typing, data is entered into a local buffer store, which is being repetitively interrogated so that this data can be displayed on the screen. Repetition rates vary from 30 to 70 cycles per second, depending on the characteristics of the phosphor and the store being used. Under reasonable lighting conditions the flicker level is not objectionable on the majority of commercially available displays.

Most terminals of this type will display up to about 1000 characters in preset matrix positions on the screen. Data which has been typed in and is being displayed on the screen, can be altered by using the edit functions, and when it is complete and correct it can be transmitted to the computer by operating the appropriate control keys. Programs held in the computer can be called into operation and data from the computer store can be displayed on the screen by operation of the appropriate sequence of control and character keys.

Various editing and control features are usually incorporated. These include the ability to insert information into and delete information from the middle of a line, with the rest of the text automatically opening or closing up to accommodate the new format. A tabulation facility would allow headings to be written up on the screen by the computer in a preprogrammed format. The operator could then fill in this form with relevant data. If a roll and scroll facility is available, text can be scanned through on the screen by regularly automatically deleting a line from the top of the screen, moving all the lines up one position and inserting a new line at the bottom of the screen.

If required, hard copy can be obtained by the incorporation in the system of a teletype, which can be triggered from the display to print out the contents of its buffer store. Alternatively, a line-printer can be incorporated local to the computer. When hard copy is required from any one of several local or remote displays, a control code from that display can cause the computer to read the contents of its buffer and print it. The line-printer output would then have to be sent to the display terminal operator.

Some terminals, such as the Cossor DIDS 400, are designed for stand alone local or remote operation. They normally contain a delay line store and all the necessary hardware character generation and control logic required for their operation. Separate models are available for "in house" and for phone line connection to the computer.

Other terminals, such as the Elliott-Automation Series 20, are designed for operation in cluster mode. A cluster share a local store, character generator and control unit. The cost per terminal in cluster mode decreases as the number of terminals in the cluster increases. In general, where more than four terminals are required on one site, they prove more economic than stand alone terminals.

Some more sophisticated systems allow the use of a lightpen (Sanders, CCI) and touchwires (Plessey). Plessey also offer colour alphanumeric displays.

Advantages of Alphanumeric C.R.T. Terminals over Teletypes

(1) They are completely silent in operation, and thus can be situated in an office, which is not dedicated to the terminals, as really must be the case with a regularly used teletype.
(2) Increased speed and ease of editing facilities on the c.r.t. terminal allow it to be used far more easily by unskilled typists—e.g. programmers, storemen, managers, etc.
(3) Vastly increased speed of output of information from the computer opens up a completely new realm of application for file interrogation. Pages of information stored in the computer can be scanned and flipped through almost as though the terminal were a book. Even if the terminal is used remotely from the computer, with the only connection being a standard switched telephone line, a dozen lines of text can be displayed in about ten seconds, over 10 times faster than a teletype.
(4) As the C.R.T. display requires a local buffer store to regularly refresh the picture, information keyed in is held in this store. This input data can be held in store until a complete message is ready for transmission to the computer. It is only then that an interrupt need be sent to the computer to receive the message. Thus c.p.u. time spent servicing interrupts is reduced. Also, if the display is being used remotely, time spent actually transmitting information over the telephone line is substantially reduced. This allows the one line to be used to service several independent displays, each being polled in turn by the c.p.u.

Applications of Alphanumeric C.R.T. Terminals

With the advent of the multi-access on-line computer, these terminals are finding application in an increasing range of fields. They are obviously of value where large data files have to be regularly scanned and updated.

Several hospitals in the United States are at present implementing computerised medical information systems, which generally process and maintain all patient records and record all activities directly involved in the care of patients. They provide a saving in physician's time as well as an extension in his ability to care for his patients.

The hardware required for this type of system includes a multi-access central processor with large disc or drum data storage files, a large magnetic tape backing store, and several remote terminals situated throughout the hospital buildings. Using one of these terminals, the physician or nurse can enter simple enquiries from the keyboard and instantly receive a reply on the screen. This reply might be patient records information, operation data, partially completed forms for insertion of new data, etc. A typical response could be a list of headings, each of which refers to a further page or pages of information. When the physician has found the information he wants, he can key in a code at the terminal and have this information printed out on a central line-printer. This hard copy could then be sent to him via the internal post.

Another large multi-access file interrogation and updating problem, which is being solved with the aid of C.R.T. terminals, is that of airline ticket reservations. Most of the major European and American airline companies either have operational, or are in the process of implementing, comprehensive multi-access systems, into and out of which the main means of data flow is through remote C.R.T. terminals. It may well soon be economically viable to use a simple alphanumeric C.R.T. display i/o system for theatre and hotel booking networks.

The use of computers for stock control is now commonly accepted practice. The ability to interrogate and update stock value quickly and easily by means of a graphic terminal has many attractions. Systems of this type are becoming quite common in the United States and are now beginning to be implemented in this country. There are many other fields where rapid access to large information files is essential. There are fairly obvious applications for this form of terminal in the banks, in the stockbroker's office and in industry.

Other major applications of these terminals depend more on their convenient input and formatting of data than on the information retrieval aspect. They are, for example, being used in the printing industry for the collation and editing of classified ad's before printing. Work is now in hand to develop hardware and software systems so that a page of text can be entered into a C.R.T. terminal, the text justified and the page layout decided and fixed. This would then act as the master page for automatic computerised print setting.

Remote Graphic Terminals

S.T.L. are developing an ingenious "Ferrodot" system[1] whereby the required diagram or text is recorded onto a white sheet of magnetic material by preferential magnetisation. The image is developed by passing the sheet through a bath of black ferromagnetic powder. This system could well be used in the future for cheap remote transmission of computer generated graphics.

The GLANCE I is an "in house" graphic display system developed at Bell Telephone Labs. using an incremental plotting technique on a C.R.T. A cluster of 16 displays are all refreshed from a central disc store, each display having its own track on the disc. The displays are used in conjunction with teletypes and have no means of receiving graphic data input.[2]

Another non-interactive remote graphic display technique is used in the BRAD System at Brookhaven National Labs. Here 10 displays are used in conjunction with teletypes as a cluster, all being refreshed from a central drum.

Graphic data is stored and transmitted in video form and the displays are standard home T.V. sets.[3]

Both of these systems rely on clustering of several terminals to achieve a cost of under $10,000 per station. Neither of these systems can be operated over phone lines as the bandwidth they require necessitates the use of coaxial cables.

Image storage tube display systems originally developed at M.I.T. are now just coming on to the market in the United States. These displays do not require regular image refreshing and thus do not need a high data flow rate from the computer. The original ARDS (Advanced Remote Display Station)[4] unit from M.I.T. has been further sophisticated and is now being marketed in the United States at $12,750 (basic) per terminal (Computer Displays Inc.). This offers alphanumeric and vector graphic input and output over a phone line with a restricted interactive capability. An interactive storage tube system[5] developed at D.E.C. relies on a cluster of displays around a small satellite processor for its low cost per terminal.

A stand alone device has been developed by Graphic Displays Ltd. in this country and is being marketed at £3,000. This is a graphic C.R.T. display terminal, which is designed to "look like" an incremental plotter at the end of a phone line. It uses a small integral drum (25 k. bits) for refresh. This allows it to display up to 300 inches of vectors or 100 quarter inch high characters.

Much work has yet to be done to ascertain the optimum trade offs between the amount of data transmitted and the terminal hardware complexity. Data compression and incremental picture drawing techniques are actively being developed at Bell Telephone Labs.[6] and at Brunel University.[7] These techniques are being used in conjunction with M.O.S. integrated circuit dynamic shift register buffer stores.

The combination of the ability to highly compress pictorial data and the likely low cost of integrated circuits, should, within the next three years, produce commercially available vector graphic remote displays for less than £2,000 per terminal. Addition of an interactive capability to these terminals will be possible without drastically increasing their cost.

Applications of this type of terminal are very wide indeed. They would find uses in practically every laboratory and office. The interactive terminals will be used for computer aided research and design and the others for alphanumeric and pictorial file interrogation.

For example, in the medical field already mentioned, there are obvious applications for the storage and display of pictorial information—for example, encephelograms, electrocardiographs, radiation isodose plots, etc.

One of these terminals on a Managing Director's desk could, with the right backing computer system, give him instant access to up-to-the minute financial, staff, work-in-progress and stock positions. It could allow him to call up computer generated shop loading graphs and compare them with graphs of work content of anticipated orders. Hardware capable of doing this is already available in the United

States and will soon be available in U.K. at fairly reasonable costs. The limitations will eventually, however, be in the costs of the software operating systems.

REFERENCES

1. Brewster, A. E., "A high-speed magnetic printout system," *Electronics and Power*, Vol. 14, February 1968.
2. Christensen, C., and Pinson, E. N., "Multi-Function Graphics for a large Computer System," AFIPS, FJCC, 1967.
3. McDonald, H. S., Ninke, W. H., and Weller, D. R., A Direct-View CRT Console for Remote Computing. International Solid-State Circuits Conference. University of Pennsylvania, February 1967.
4. Stotz, R. H., A New Display Terminal, *Computer Design*, April 1968.
5. Ruben, M. A., Low Cost Graphics. International Symposium on Computer Graphics. Brunel University, July 1968—see p. 91.
6. Ninke, W. H., Inexpensive General Purpose Remote Display Terminals. International Symposium in Computer Graphics—Specialist Session, Brunel University, August 1968. (Private communication.)
7. Pitteway, M. L. V., A Simple Data Compression Technique for Graphic Displays or Incremental Plotters. International Symposium on Computer Graphics—Specialist Session, Brunel University, August 1968. (Private communication.)

Biographical Note

R. Elliot Green read Applied Physics for a B.Tech. at Brunel University. He worked in the Nucleonics Division of E.M.I. Electronics Ltd., until 1967, when he left to join CEIR Ltd. The Company have now changed their name to Scientific Control Systems Ltd., and he is their consultant specialising on all aspects of computer graphics. He is also engaged in research with the Department of Computer Science at Brunel.

PART 2
Applications, Installations

Graphical Computer Aided Programming Systems*

C. W. GEAR

University of Illinois

Introduction

By the term "Graphical Computer Aided Programming Systems" we mean those systems in which the computer plays a fundamental role in the preparation of a program as well as being the executor of the end product. Interactive, line at a time, compiler systems fall into this category. The graphical systems that we are going to discuss use some form of flowcharts as a part of a program or its documentation. Flowcharts may be associated with the four phases of program preparation:

> Planning
> Coding
> Debugging
> and Documentation.

Various ways in which flowcharts have been used will be surveyed below, and three particular projects which involve an on line interactive use of a graphics terminal will be discussed in more detail. These projects are

(1) The GRAIL[3] project at the RAND Corporation
(2) W. R. Sutherland's[11] work at M.I.T. Lincoln Laboratory
(3) The work of F. K. Richardson, T. Y. Lo and the author[8] at the University of Illinois Department of Computer Science

Naturally, the third will receive the greatest attention because we are the most familiar with it!

The Programming Process

The processes involved in the preparation of a job for computer solution are shown in Figure 1. The initial stage in box 1 is the job breakdown. On a large job, this is not done by the eventual programmer, but at the programming management level or by the system analyst. No computer is used at this stage. Box 2 shows the testing of the initial breakdown to see if it is feasible. This step is not usually performed now, but left until the final step in box 7. Worse than that, because the manager has performed step 1 and the programmer tests it in step 7, it is a simple rule of promotion that there are no errors in step 1. Consequently changes will be made at the wrong level, resulting in the software delays and inefficiencies of which we are all too aware.

---

* Supported in part by Contract No. AT(11-1)-1469 of the Atomic Energy Commission and in part by the University of Illinois.

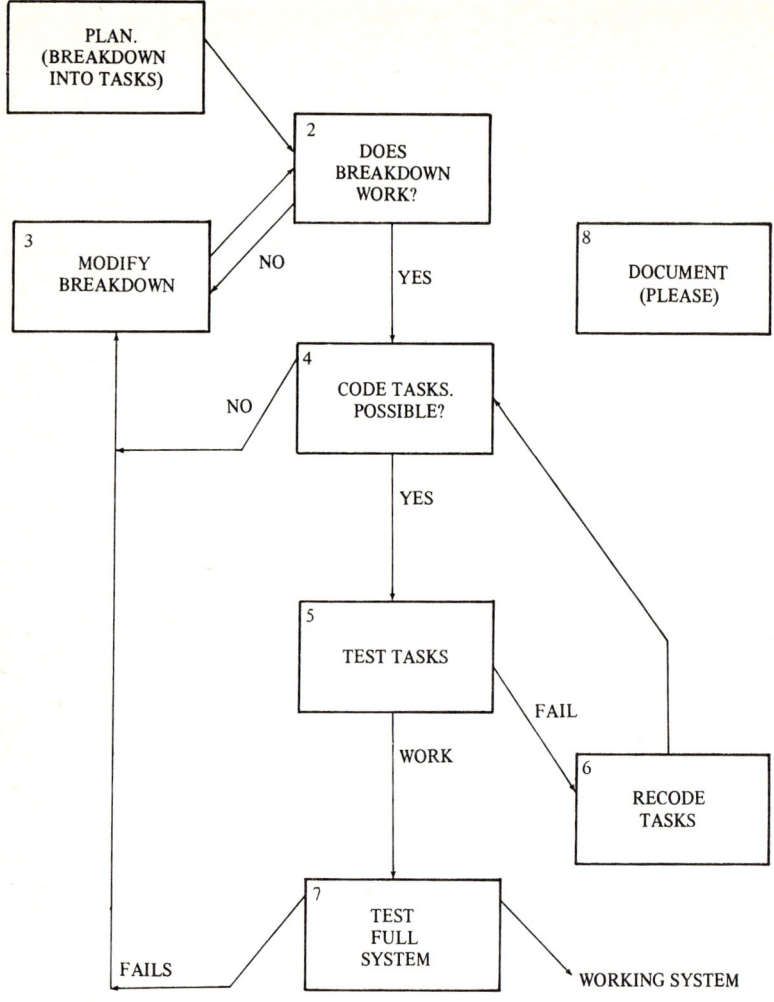

Fig. 1   Program preparation

The tedium of returning to modify the overall concept of the job often forces the programmer on a one-man job into costly "patches" of his program at later steps. It is therefore desirable that some amount of testing be applied to the results of box 1. These may result in a modification of the plan.

When the plan is checked out, the programming of the individual tasks can begin. At this stage it may appear that the plan is unreasonable. Interfaces between different tasks may be incomplete. It may not be possible to implement one of the tasks, either because of space and time allowed or because of a logical contradiction in the specification.

A number of projects have run into trouble because a failure to meet the space and time requirements has not been allowed to cause a return to the re-planning step 3. When the tasks have been programmed, they are tested on the computer. Since this is usually the first approach to the computer, the individual tasks are tested at step 5 in a simulated environment. Each programmer will prepare a set of test programs that, he hopes, will exercise the task in the manner he believes the total system requires. Were the overall plan and task specifications available to the computer at this time, a realistic environment could be created for testing each task, and the amount of time used in the preparation of tests reduced. As a result of the tests, some of the tasks will have to be reprogrammed, possibly leading to a failure to meet specifications in box 4. This also can lead to a modification in the original breakdown. In box 7, the total system must be tested. With present methods, this is the only time that the original plan is tested, and hence may lead to more modifications to the original plan than necessary.

Finally, in box 8, the documentation is produced. This is not part of the regular flow, but should be produced in parallel with the production of the plans and tasks. Modifications are often inadequately documented because the supervision of documentation is not a computer function.

The Use of Flowcharts

Flowcharts have been coupled to the programming process in a number of ways. Attempts to solve the documentation problem (the fact that it is not an integral part of the programming process as shown in Figure 1) have been made by providing programs which convert automatically, or semi-automatically, from a program prepared in a conventional form such as in FORTRAN or assembly language into a flowchart. Most of these do little more than replace conditional and unconditional branches with arrows to the appropriate section of code contained in a box. Little or no attempt is made to group parts of the program that are logically "close".

Knuth[7] requires that the logical flow be specified in the first few comment columns, and from this draws a "linear" flowchart, that is, there is a single downward flow of flowchart boxes so that the second dimension is not used except for arrows going up or down. Sherman[9] analyses the flow automatically, and can specify the source (input) language by means of a meta-language that can handle languages without a block structure (e.g. FORTRAN and assembly language but not ALGOL or PL/1). A copyright program Autoflow,[1] which is available commercially, performs a similar job, producing flowcharts with up to four columns of boxes which serve to reduce some arrow lengths. No detailed description of their method appears to be available in the open literature.

Two somewhat different efforts have been made related to the logical connectedness of the program. An early one (1959) by Haibt[4] examines the total flow to try and detect self contained groups of code with few input or output connections. This could provide a basis for automatically dividing a program into subroutines as well as provide a breakdown into sections that should be printed on the same page. Hain and Hain[5] discuss the problem of the automatic layout of flowcharts on paper to reduce arrow lengths. They do this by breaking connections of the associated graph until it is a tree. The flowchart can then be drawn without backward arrows. Connectors are used to indicate backward connections.

These techniques mainly serve to aid in the final documentation. If the programmer has used adequate comments in the source program, the final flowchart may provide a more "visible" presentation of the code. Debugging has also been aided by these methods (if a new flowchart can be generated reasonably quickly) but the more sophisticated system is usually less useful for debugging. This occurs because the simple system that attempts no automatic layout only changes the visual presentation in those areas in which the program is changed. A sophisticated layout procedure may well modify the whole flowchart in response to a change of one "GO TO" statement. (Part of the problem of documentation can be solved by avoiding the "GO TO" statement—see the letter of Dijkstra.[2] However, there are many situations in which the overhead of the alternative approach of recursive procedures is currently unacceptable.) The programmer remembers much of his program "visually" in terms of its appearance on the printed page, so the changes to the layout in the debugging process should be minimised.

The techniques discussed above were output oriented—that is, they relied on output devices to produce better documentation. Input orientated techniques involve the preparation of flowcharts as the source for the program. In a batch processing environment a version of this flowchart must be passed to the computer. This could be done with a sophisticated optical input device, but this is probably not practical and hasn't been attempted to this writer's knowledge. Alternatively, the flowchart must first be converted to a more conventional card orientated language. This has been done by manual transliteration in some documenting systems (undocumented as far as we know!) and is also being done by a subsidiary graphic computer at the University of Illinois. In this system, which is a subset of the system to be described later, flowcharts are drawn on the face of a C.R.T. and can be converted to conventional languages such as FORTRAN when required. The information entered in the flowchart boxes must be legitimate statements in the object language. Decision boxes and arrows transferring into sections of earlier code are replaced by appropriate control transfers specified in a meta-language description of the object language. This system provides some final documentation the moment that the program is finished by ensuring that the flowchart is always a true representation of the program as well as a useful debugging tool. It is hampered by the fact that the results of the program execution are related directly to the object program rather than to the source flowchart, and its economics are dubious because it requires a small graphics system (the DEC 338, about $100,000 for the particular system with DECtape and 16K of core). This is more expensive than a keypunch. The DECtape drive makes it possible to retain the source on magnetic tape at all times, and edit back onto tape. If the potential reduction in the cost of systems similar to this were coupled with an ability to read low cost magnetic tape storage directly into the batch computer, it could prove to be a useful keypunch replacement for productive programmers.

The major benefits of flowchart input and output arise when they are used in an interaction system. The GRAIL[3] project and W. R. Sutherland[11] used a C.R.T. display and tablet/light pen input coupled directly to a central computer. At the University of Illinois we used a DEC 338 as a remote graphics terminal coupled over a 100 character/second line to the ILLIAC II. (It has since been coupled also to an IBM 360.) The GRAIL project has been primarily interested in understanding and broadening the relationship of the man and the machine he is working with, by limiting communication to the graphic I/O. They have one button which initialises the system. The computer performs on line real-time recognition of handwriting and lets the user draw flowcharts or assembly language programs. Several different box shapes are recognised and replaced by a computer drawn version. (The beauty of on line recognition of handwriting is that the human can immediately correct errors when the computer redraws a character that it has recognised. Currently the amount of processing required does not appear to make this an economic input medium for general use, but I think that we can expect the cost of hardware and the length of the algorithms to decrease very shortly to the point that such a system is attractive.) A flowchart box may represent a full flowchart or a section of 360 assembly language code. Sutherland used sections of Sketchpad[10] to allow him to draw arbitrarily shaped symbols, each of which represented an action. Some primitive actions such as +, *, =, etc., were defined, so that the shape of the flowchart symbol contained the semantics of that symbol.

A conventional flowchart uses lines between boxes to represent control flow. Sutherland also used them to represent data flow more in the manner in which a flowchart is drawn for an analog computer. Thus the ADD primitive required two inputs and had one output equal to their sum. It was activated each time that either input changed, although it was possible to require either one or both inputs to change, or to use control flow. This technique allowed parallelism to be indicated implicitly. The use of lines for data flow solves the problem of naming variables and the question of the scope of names. This is usually handled by a block structure in a conventional language, but that is not always the most desirable method, and the analog of block structure is somewhat artificial on flowchart.

The effort at the University of Illinois has been principally concerned with the implementation of a flowchart programming system on a remote graphics terminal in a manner which may prove to be an economic tool in the future, and such that the computer is brought into the process as early as possible. The remainder of this paper will describe the system briefly.

Present Status of Systems

At the main machine level, an interpreter has been written for a number of basic machine functions. These are called primitives. They include arithmetic capabilities. Additional functions can be defined by means of flowcharts. These are essentially added to the set of available operations, so that there is no distinction, to the user, between those initially provided and those that he defines to simplify his programming job. For hard copy documentation, it is possible to get a CalComp output copy of a flowchart. At the terminal, the user can write and update arbitrary flowcharts. These flowcharts can be executed by calling for them at the terminal and feeding them data. If desired, a flowchart can be traced during execution a box at a time so that the logical flow of the program can be observed. Figure 2 shows a simple flowchart which defines a new function POLY as a piecewise polynomial function of its argument X. For negative X, POLY has the value 3, for positive X it has the value $3 - 5X + 4X^2$. The first box defines the name of the function, the second sets its argument X. The third tests the sign of X, while the remaining boxes calculate the value of the function for return to the calling program. This flowchart will define POLY so that it can be used as a pre-fix operator in other flowcharts. The language currently used for typing a program into the boxes, was developed in the department and resembles the language SPRINT.[6]

We shall not examine the language in detail, but any standard programming language could be used in a flowchart programming system with minor modifications so that control flow is represented by the flowchart structure.

Once a flowchart has been constructed at the remote terminal DEC 338 independently of the Illiac II, it can be saved for later use or immediately tested. It is saved on local back-up storage (DECtape) if desired. It can be retrieved at any time and edited on the remote terminal.

When a flowchart has been written or edited, it can be transmitted to the ILLIAC II as sequences of lines of characters in the format prescribed for the ILLIAC II time-sharing system. These lines contain information about the topological, positional, and textual information. Once the flowchart has been transmitted to ILLIAC II it may be executed and the flow of execution observed at the terminal by the blinking of

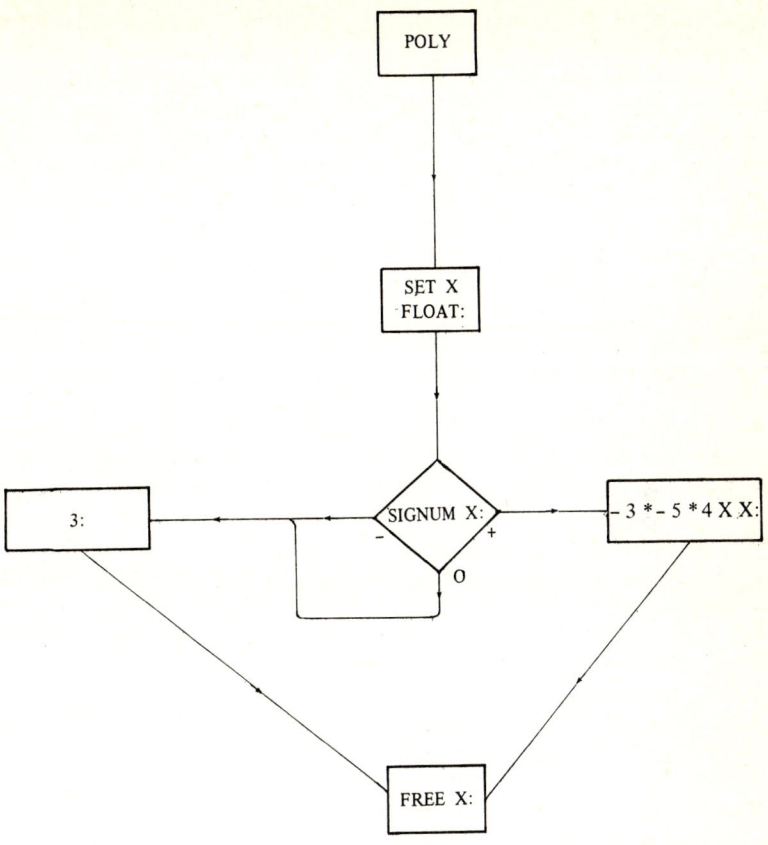

Fig. 2 POLY routine

Underline indicates output from the ILLIAC II. Comments in lower case.
LOAD: transmit the flowchart poly to the ILLIAC II
SETIME 2: set the time in seconds to be taken between executing boxes
MTRACE < POLY > : trace the flowchart poly
= GRAPH POLY 1 : PRINT GRAPH: execute poly with the argument 1,
                                          store result in graph
GRAPH =2.0000000000000   printed value of graph from ILLIAC II
= GRAPH POLY 2 : PRINT GRAPH: the argument is 2
GRAPH =9.0000000000000
= GRAPH POLY 0:PRINT GRAPH: the argument is 0
GRAPH =3.0000000000000
= GRAPH POLY − 0 2:PRINT GRAPH: the argument −2
GRAPH =3.0000000000000

Fig. 3 Text Communication with ILLIAC II

the symbol currently being executed. The execution of the flowchart can be stopped at any point, and the values of any named variables examined. The process of execution is illustrated in Figure 3. The EXEC mode is entered by means of a light-pen button and the command LOAD is typed. This causes the flowchart to be transmitted to the ILLIAC II. From now on statements in the language interpreted by the ILLIAC II can be typed directly from the terminal to the ILLIAC. A number of these are illustrated in Figure 3. The flowchart POLY is executed and traced visually in order to check it.

What Next?

We have seen that it is possible to get additional documentation by bringing the computer into the picture at an earlier stage. What else can be done to speed the preparation and modification of programs, and in what way can the facilities provided by a graphic remote terminal be used to best advantage? It is obvious that we should provide for direct graphic input-output within the language so that those problems in which graphic I/O is crucial can be handled. The extent of machine checking of the relationship between the plan, that is, the highest level program, and the tasks, that is, the lower level programs, should be improved, and the requirements for a full statement by the programmer should be made as stringent as possible so that documentation is more complete. The problems of multiple terminals working on the same job need to be investigated in detail. With such a system, it would be possible for the manager and his programmers to communicate through the machine, with the machine checking that the programs are compatible, providing the plan to each programmer, and informing the manager when it must be modified.

Finally, the use of flowcharts as a source language makes the statement of parallel tasks simple. (Notice that in Figure 1, we specified the parallel occurrences of documentation with the rest of the serial steps.) One such extension has been performed by W. R. Sutherland.[11] Most programmers, we suspect, think in terms of sequential operation, probably because all present day programmers have been trained to this end. We don't think that the non-programmer typically thinks in that way. There is a need for an adequate way to express the inherent parallelism of a job, and we believe that flowcharts will provide this.

BIBLIOGRAPHY

1. "Autoflow," A program leased from Applied Data Research, Inc.
2. Dijkstra, E. W., Letter to the Editor, Comm. ACM, Vol. 11, No. 3 (March 1967), pp. 147–148.
3. Ellis, T. O., Sibley, W. L., "On the Problem of Directness in Computer Graphics," Proceedings of "Emerging Concepts in Computer Graphics," Fall 1967. Edited by Secrest and Nievergelt (W. A. Benjamin) pp. 123–130.
4. Haibt, L. A., "A Program to Draw Multi-Level Flow-Charts." Proceedings Western Joint Computer Conference 1959, pp. 131–137.

5. Hain, G., Hain, K., "Automatic Flow Chart Design," Proceedings 20th Nat. ACM Conference, 1965, pp. 513–523.
6. Kapps, C. A., "SPRINT: A Direct Approach to List Processing Languages," SJCC Proceedings, 1967, pp. 677–683.
7. Knuth, D. E., "Computer-Drawn Flowcharts," Comm. ACM, Vol. 6, No. 9, pp. 555–563. (September 1963.)
8. Richardson, F. K., Lo, T-Y., Gear, C. W., "Computer Aided Programming System," Proceedings of "Emerging Concepts in Computer Graphics", Fall 1967. Edited by Secrest and Nievergelt (W. A. Benjamin) pp. 171–184.
9. Sherman, P. M., "FLOWTRACT—A Computer Program for Flowcharting Programs," Comm. ACM, Vol. 9, No. 12, pp. 845–854. (December 1966.)
10. Sutherland, I. E., "Sketchpad: A Man-Machine Graphical Communication System," Tech. Report 296, MIT Lincoln Lab. (1963).
11. Sutherland, W. R., "On Line Graphical Specification of Computer Procedures," Report ESD-TR-66-211, MIT Lincoln Lab. (May 1966).

Biographical Note

C. W. Gear received a B.A. (Hons.) from Cambridge University in 1956 and the M.A. in 1960. He also received the M.S. from the University of Illinois in 1957 and the Ph.D. in 1960. From 1960–62 he was with IBM, England. He is at present Professor of Applied Mathematics in Computer Science at the University of Illinois.

Alphanumeric Terminals for Management Information

A. E. P. FITZ
Ministry of Defence

This chapter describes the use of alphanumeric terminals for management information in two Ministry of Defence computer installations and the first reactions to their operation.

We have two technical stores depots, which are more than just storage sites. Their staff is engaged in buying, in forward planning, and in all the work involved from taking over from R & D until the time that goods are in store when there are the customers to be supplied. They are the nearest things that one would get in a government sense to business systems and in this respect they differ from most of the scientifically oriented systems which people have been describing up to now.

The computers in these depots are Univac 492s, with magnetic drums capable of holding approximately 400 million characters of records, and eight million characters of programs and ancillary storage. Thus, on direct access drum records, are held all the current records of the business, such as the stock records, parts explosions, financial records and so on. All that goes on gets logged onto the drums of the computer. The design of the system is to grind continually at ordinary business tasks, e.g. invoicing and forecasting, but at a level above the steady grind to have priority interrupt work which in general jargon is known as real time, and part of this is enquiry work, which is the subject of this paper. The terminals which we use are Cossor DIDS 400, grouped in a cluster of 5 connected to a control unit. This leads to economy, since the one control unit can take up to 64 terminals all operating independently, and this does cut down the price if one is planning for more terminals.

The terminals themselves are sited in various offices of the business, they are accessible to anyone in those offices, and they can display all the records which are held on the drums. No attempt has been made to limit access to specified people, and all files are accessible. However, the suite of enquiry programs is activated only during normal office hours, thus preventing the use of the terminals by unauthorised persons.

A series of numbered questions which cover all fields of the records have been specified so that by quoting a file number which anyone will be familiar with because it is a number he uses every day, and by quoting a question number from a handbook provided with the terminal, one can bring up on the screen, a selected part of any file, and look at it. This is a passive system and there is no feedback. If one wants to know, e.g. the holding of any item, one types in the file number, the enquiry number, presses the transmit key, and the answer comes up on the screen.

What happens when the transmit key on the terminal is pressed, is that the query is sent into the computer core store via the control unit. In core it is in ASCII characters, and these are translated at once into internal character format, and are stacked out of core on to a drum. This drum staging is necessary, because one cannot predict the rate at which enquiries will be made and we wish to restrict the amount of core used for enquiries because these might tend to exclude routine work which must go on.

At the same time as the message goes on to drum, an entry is made in a task queue which the computer works through item by item, bringing in the application program to deal with enquiries. This program has its own buffer areas. It calls down the next message from drum; it finds out what file is required; it calls a priority interrupt for the file to be brought in; it picks out the fields it wants; it labels them (since the only information one has in the actual fields is numeric information, this must be prefixed with a description to make it meaningful on the screen); the application program prepares the message for output, stacks it up on the drum again, and gets out of the core. Again a queue is entered once the message is up on the drum, waiting to come down to the core again for translation into ASCII, and transmission out to line to the enquiry terminal (which, of course, has prefixed all input messages with its own address so that the answer comes back to the right place).

That, very briefly, is what happens, and although the actual details are a bit more complicated, in practice the interval between transmitting an enquiry and receiving an answer, seldom exceeds a second or two.

A question to be asked is, "Was the cost of installing these terminals worthwhile?" In fact, it didn't really cost us very much more to install these terminals than to leave them out, because we had originally designed a system where all the current records were held on drums. We had had to write a file handler so designed that the applications programmer says, "Get File N" and the software brings it into core, so we already had a process for bringing records in. We already had programs for polling and queueing written, to deal with other priority jobs, so it was easy to add a communications routine on to the poll, which made use of the facilities already designed, to send out portions of the files on to the screens. We did have to decide the extent to which we'd let people use these screens. I made the decision (and I frequently have to restrict bright ideas) that it was sufficient to let people use what existed, without trying to be too clever too early. I consider this important, because people don't want to be promised something tomorrow, but they want something today. If one can get something simple going quickly, it has a very much greater psychological effect than the advanced plans for the future which are so often presented. The effect of using these terminals was splendid because one thing I am continually striving against in our installations is a "We and They" complex. One goes and buys a computer from a manufacturer, and he spends a great deal of money giving one the hardware and software, and a lot more money in educating computer people. He picks out suitable staff and gives them systems courses and programming courses, but nobody thinks of the people outside the computer complex. It's not really the manufacturer's job, it is the management's job to see that they get the education, but they very often don't get it, and they regard those in the computer centre with a certain amount of resentment, feeling they've been misused. This is a communications, or participation problem, and difficult to overcome, but give them a "goggle-box", and say "You can look at your records: they are in there, they don't belong to somebody else, they belong to you", and they can use their terminals, and see those records which in a total system are up to date, and factual. A man doesn't have to ring down to the stores to ask for details of yesterday's stock position or ring the finance department to see how much cash is available in a budget: he can find out for himself and in fact there has been a significant reduction in internal telephone traffic. This is one of the tradeoffs: It gives a sense of participation to those who have access, and this then gives a second tradeoff: where these terminals have been put in, people do participate. Instead of having misgivings about the implementation of a computer system, they are pushing the teams to get on with the work, and this is the first time I've met staff who want to give work to business computers. This situation brought about by the use of these terminals occurs because there is now participation between those who compute and those who do not directly compute. The problem may not arise so much in scientific work where it is easier to get to a computer and do some programming oneself, but in business this is not possible, and the use of terminals is therefore very worthwhile.

Used simply as enquiry units, they have proved, during the few months they have been in operation, to be easily operated, completely quiet, flicker-free and reliable mechanisms of undoubted benefit and these early results have encouraged extension of the system to remotely sited units, operated via modems and a direct telephone line, for use by industrial supervisors in the storage areas.

Regarding the use of Teletypes, we have them and they are much cheaper than video terminals, but until the two are operated side by side, one does not appreciate their difference in speed. Exactly the same basic programs work for both sorts of terminals, but for the same questions put in, one sees how much longer it takes to type out ten lines of information than to have them flashed up on the screen. One might contend that it is unnecessary to have ten lines, why not just one? This is possible, but restrictive with so many records on line and in any case we find one seldom asks a single question. If one does not get the answer one wants at first, it may be necessary to ask two or three, and what takes seconds on the screen, can take minutes on the Teletype. Herein lies the difference, and for the people actually using the terminals, there is a quickening of the pace of work deriving from their ability to solve everyday problems more quickly, and a resulting general sense of job satisfaction. Briefly then, I think these terminals create within a business, an *esprit de corps* which you do not get in any other way, and this is worth a lot.

One further point is development. There have been various suggestions that, e.g. a managing director might have a terminal on his desk. This I personally am not in favour of. The top man's work is essentially in long-term projections. He doesn't want to be breathing down people's necks, and it is the job of people further down to keep in touch with the actual details minute by minute.

Another point is how far do you go in putting in data by means of these terminals? This is something I am very chary about. If there is a good system, with all sorts of checks against errors, and if people are allowed to alter records via terminals, then all the defences built up in the machine tend to fall flat. The next stage from passive enquiries as far as we are concerned is "time regulated actions". For instance, suppose somebody rings up and asks if an item is available in the stores, and it turns out to be the only one in stock. Since there is work going on all the time by the time a properly documented entry is made, someone else may have taken it out. So there is no reason why the enquirer should not be able to put a thirty minute earmark on the item. After thirty minutes, the clock in the computer will cancel the entry, unless a proper application is made. This application requires extensive vetting against a check list, and it is not practicable to do it on line with people who are essentially only "two-finger" typists. It is better to let them reserve items, and then have the input data prepared in punched card form so that the computer has sufficient information to prepare correct invoices, and do the other statistical analyses resultant on the orders. Beyond that I'm not going to go: I try to get people to make small returns quickly. So for those who, for example, are wondering how to allow doctors to feed information in through this sort of terminal, I would say "Don't". There is plenty of benefit to be derived now with simple programs which can greatly improve information flow, and hence morale. If one can get something going quickly which gets the staff participating, this is worth a lot of bright ideas for tomorrow. There is never a shortage of plans and proposals and it's very attractive to be one of a team continually way ahead of the rest. But not too far out, please. Evolution, not revolution, is what I like to encourage.

Biographical Note

Mr. Fitz is a Senior Supply and Transport Officer in the Navy Department and has been a Civil Servant in the service of the Royal Navy for over 30 years. He has had an interest in data processing since his pre-war years in Chatham Dockyard which at that time had a single Hollerith installation. In recent years he has been associated with the development of computing facilities for the Polaris project and at present supervises all the computer facilities in the Stores department.

Computer Graphics Used for Architectural Design and Costing

P. E. WALTER
West Sussex County Council

Proposition

The West Sussex County Council has been using various computer applications for a number of years. Under the direction of Bernard Peters, the County Architect, the Department of Architecture has been accelerating the evolution of computer techniques to assist the designer and his colleagues in the building team since it first produced a bill of quantities programme about two years ago.

The proposition to add a graphic display unit to the existing computer configuration to further assist the building team had to be economically viable and the techniques described in this paper recognise this fact and have given to the computer a very large share of the work load in the shortest possible time, at the same time leaving the Architect his prerogatives of design and aesthetic evaluation. They also allow the continuing expansion of the basic programme so that it may cover a much wider field than it does at present. The major, if not the only limiting factor on the rate of this expansion is that of manpower resources available.

Examination

It was very apparent from the commencement of this project that the procedures and processes which must take place between the time that the client commissions the architect to design a building and the time when the completed building project is handed to the client and the final account settled would have to be rationalised and many traditional thoughts and methods would have to be abandoned in favour of new techniques.

The areas of the time scale with which this chapter is concerned are the design and costing of the building project.

It was argued that if a building project was considered as a conglomerate whole which could be broken into sections each of which could be sub-divided and if each sub-division could be further sub-divided and so on and so on a chain would have been established from a complete building to its smallest component part. In this way relationships between parts are established, but at the same time each individual component can be uniquely indentified.

Upon examination it was seen that a building consists of a number of major components each one of which performs a different function. For example there is an envelope which encloses a volume of space. Within this volume of space dividers are used to create smaller areas.

The atmosphere within these areas, temperature, humidity, level of lighting, etc., have to be controlled so that the occupants are provided with the most suitable conditions in which to carry out their activities. These activities may require items of furniture, machinery, sporting equipment, washing facilities, etc. to be provided.

Solution

In identifying these functions a list of 11 groups was established which together would form a complete building. The complete list of functional groups thus established is:

(a) site
(b) structural support
(c) vertical envelope
(d) internal space dividers
(e) base platform
(f) intermediate platform
(g) horizontal envlope
(h) servicing
(i) vertical circulation
(j) environmental control
(k) external environment

A strict discipline is applied to the functional groups which is that they must occupy their own station or zone and must not trespass on any area designated to another functional group. Neither of course must they be trespassed upon. Figure 1 shows the division of part of a building into six of the functional groups.

Next, each functional group was itself divided and this time the discipline applied was that of three dimensional shape. These shapes we call "blobs" and are defined as three dimensional volumes of space. Using a four inch module, the minimum number of different "blob" shapes were carefully designed for each functional group. Using a combination of these "blobs" the designer can create plan shapes, with a 12 in. planning discipline, suitable for the types of buildings for which the authority is responsible.

These "blobs" can be filled or partly filled with material or indeed need have no content at all.

Construction

Decisions regarding the type of material with which to fill these "blobs" and measuring the quantity of material in the "blob" shape quickly produced a library of units of materials from which the designer can choose. Figure 2 shows the "blob" shapes devised for internal space dividers.

From the libraries of "blob" shapes and the materials that may be used to infill them it has been possible to produce catalogues each of which contain "blob" shapes and infillings applicable to (a) a particular type of building; for instance old people's homes, schools, fire stations, libraries, etc., or (b) a particular system of building, for example Concrete frame, S.C.O.L.A. or Rationalised traditional building. Thus the catalogue prepared for Rationalised traditional building would probably include "blobs" of different shape from that provided for a

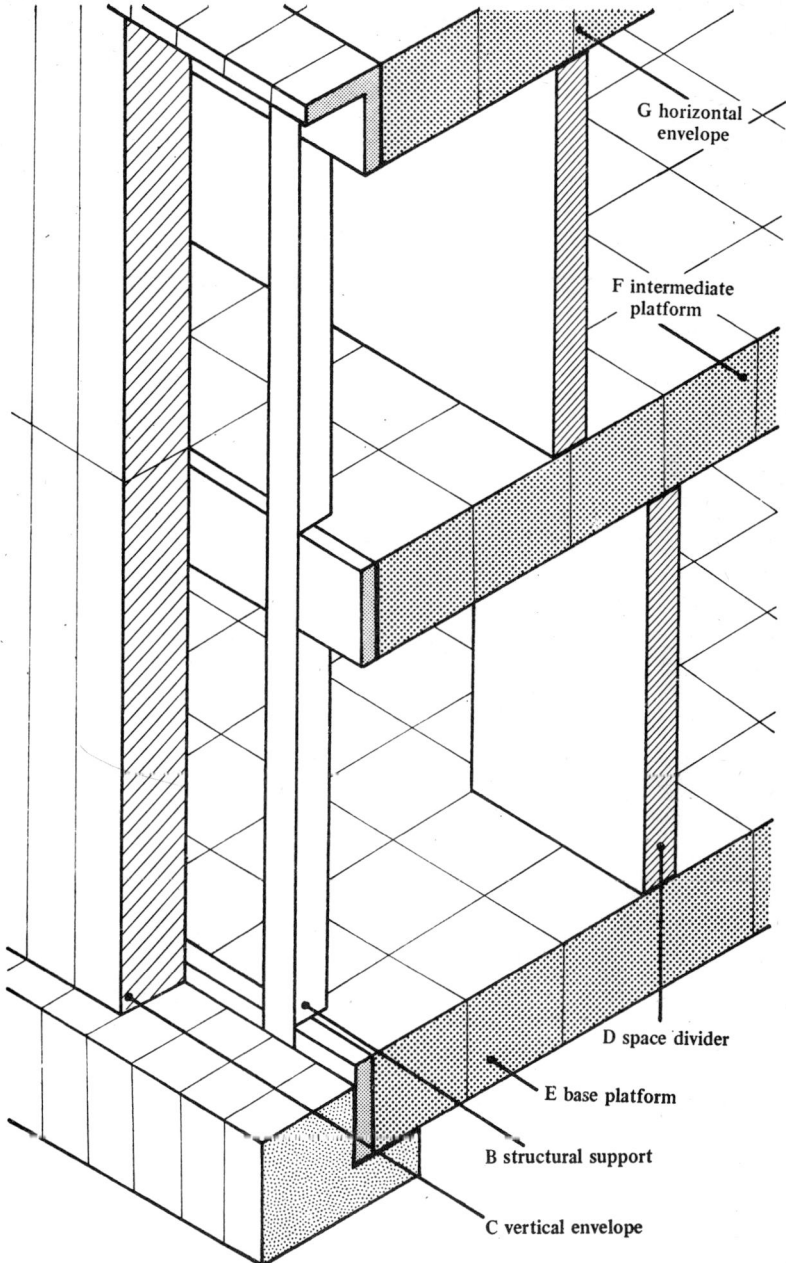

Fig. 1 Division of a building into functional groups.

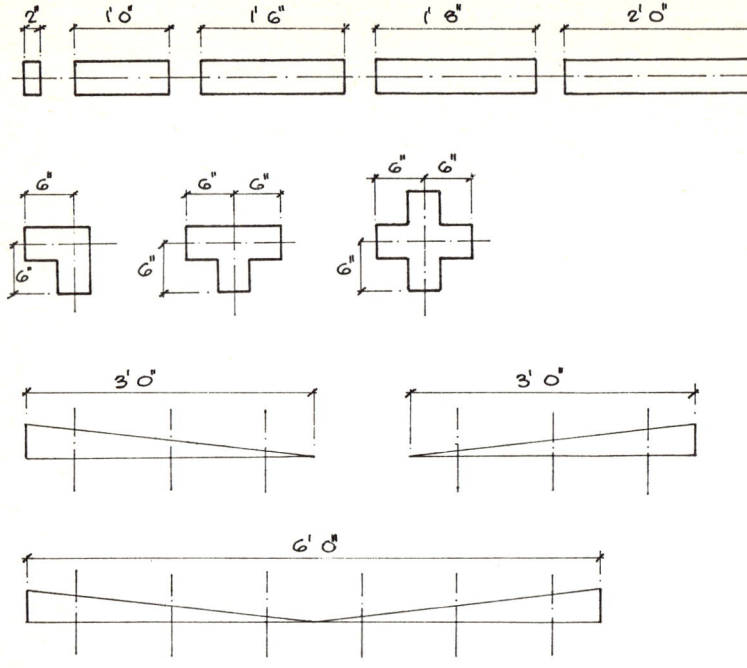

Fig. 2 "Blob" shapes for internal dividers in buildings

concrete framed building. The designer having chosen the type of system building he is going to employ for the particular type of building he is going to design, selects the appropriate catalogue knowing that he will find therein the materials most suitable to this type of building project. He will know that the materials comply with any requirements that client or statutory regulations might enforce. He will know also that the most up to date materials and methods have been investigated and decisions made as to their suitability for the type of building concerned. These checks having been made before the unit of material is allowed to be included in the catalogue, frees the designer from having to do a large amount of research which is time consuming, often repetitive and where normal procedures are followed frequently means the use of designer's own limited vocabulary of materials because of lack of design time.

The catalogues will later be supplemented by a computer based enquiry service currently being designed which will produce for the designer the materials which comply with performance standards that he specifies on a question and answer basis. For example the designer having informed the computer through a tele-processing terminal that he requires a "blob" infill for functional group (c) (vertical envelope), will be offered a set of performances. That is weight, thermal insulation, sound transmission, etc. against which he will nominate one of a specified range of values that he wishes the material to provide. The computer will respond either with material or a short list of materials complying with the specification or a request for more explicit re-specification.

The designer having selected, functional group by functional group, the units of materials that he wishes to use, makes known his selection to the computer and identifies it with a job number. Each time a different selection is made a unique code is created consisting of the functional group, the infill material and the "blob" shape. Thus every infilled "blob" identified with a particular job carries its own code.

Operation

Ideally the designer should now go to the graphic display unit and commence designing his building using the light pen. However it must be appreciated that a lot of design time is thinking time and until we have more sophisticated computers which are cheaper to run and until more designers have been trained in the use of these techniques then the designer cannot be allowed to come to the display unit until very much later in the design process. A reasonably hardened plan shape must therefore be produced by the designer before the display unit is used.

The computer configuration that we have in West Sussex is an IBM 360/40 system operating under O.S. The 256 K. central processing unit is partitioned, making 90 K. available for our program. A card reader, card punch, line printer, four tape units, five disc drives, one data cell, 2250 model 1 graphic display unit and a number of tele-processing terminals complete the configuration.

The display unit itself comprises a cathode ray tube equivalent to a 19 in. television screen, a light pen, which is a photo-electric device activated by a foot switch, an alphanumeric keyboard and a function keyboard having 32 keys which can be overlaid eight times.

Having come to the display unit with the reasonably hardened plan the operator activates the 2250 and a message is displayed on the uppermost part of the screen asking the operator to specify the number of the job that he is going to work upon. This specification is made by the use of the alphanumeric keyboard and the operator is advised by another message when the computer is ready for him to commence working.

A permanent display of screen symbols provides the operator with the facilities to rotate and slide symbols, page the various menus of units and symbols and indicate special details not included in the system. Having received the message that the computer is ready, the operator by depressing one of the function keys calls for a grid of dots to be displayed in the work area which will give scale to the drawing.

The depression of another function key produces a display in a menu area to the left of the screen, the list of functional groups coded by the letters A through K. The operator then indicates, by use of the light pen, the functional group with which he is first going to work and in place of the menu of functional groups he now has displayed for him a menu of the units of materials that the designer elected to use within this functional group. He now selects one of the units with his light pen and the computer responds by displaying along the bottom of the screen a menu of "blob" shapes which are available to him and associated with the unit of material that he has selected. Using his light pen he now points to the blob shape that he wishes to employ and then to the point in the work area where he wishes it to be placed. By the process of first pointing to the "blob" and then the position in the work area he composes the plan. Sequential light pen detects in the work area will produce repetitively the same "blob" shape.

Because there is insufficient space in each "blob" shape to include the full nine digit code attached to it, a temporary single alpha representative code is applied job by job and is used on a "coding by exception" basis. That is to say, in preliminary documentation it is made known to the general contractor (who will receive copies of the screen image as explained later) that unless otherwise shown, all uncoded "blobs" in a particular functional group are as specified but the temporary codes "A", "B" or "C", etc., shown as this job represent . . . and the full unit code is quoted.

Also provided are facilities (*a*) to delete "blobs" from the work area; (*b*) to file on disc a completed picture for subsequent retrieval if required, and (*c*) as a guide to the operator there is displayed for him the computer assigned number given to the picture and a running total of the number of "blobs" he has used in the composition of the picture. The operator is also notified when the number of "blobs" has reached the capacity of the machine.

Production

Obviously some check must be put on the cost of the building to ensure that it does not exceed the cost limits which are applied to it and this is achieved by involving a technique which the authority has been using for about three to four years called serial tendering. Contractors are invited to compete for a block of several projects, the geographical location, approximate cost and estimated start and completion dates of which are made known to them and they are supplied with a list of activities against which they put their pirices for executing. For example they might be asked to quote a price for building a square yard of 11 inch cavity wall or laying a square yard of tarmacadam paving or glazing an area of window.

These prices are then run against a dummy bill of quantities and the price of a fictitious building is then calculated. The results thus achieved influence the decision as to which contractor shall be awarded the serial.

The rates quoted by the successful contractor can now be related to the units of material which infill the "blob" shapes and thus every time a

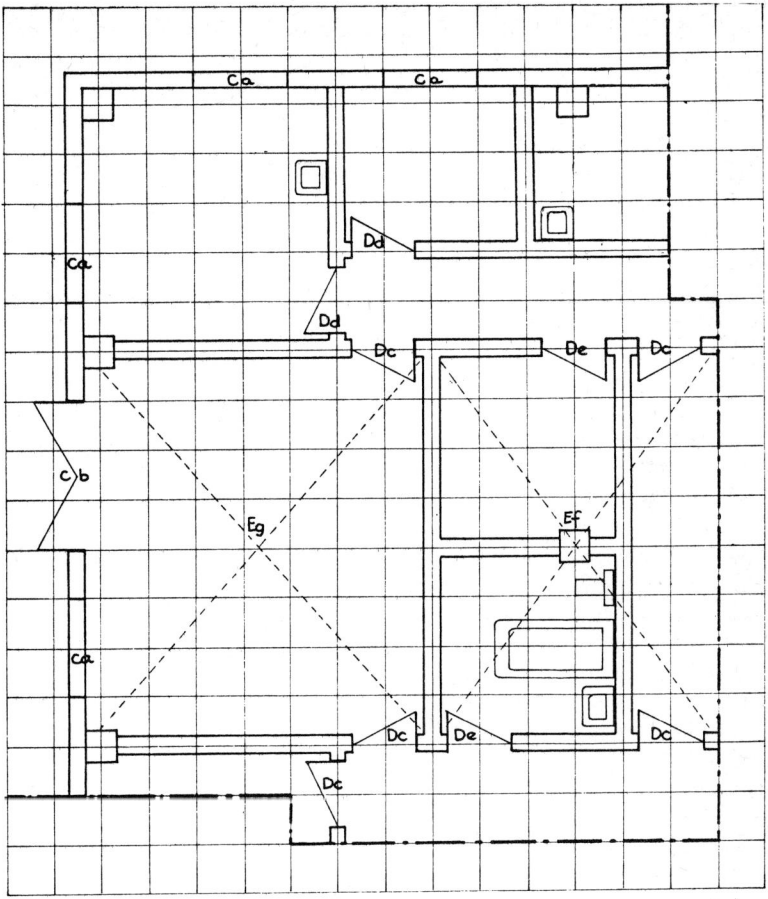

Functional Group		
C	**VERTICAL ENVELOPE**	
	All of the vertical envelope to be code	001/A
	except where marked a which will be	001
	b	002
D	**INTERNAL SPACE DIVIDERS**	
	All space dividers to be code	001
	except where marked c which will be	004/A
	d	004/B
	e	002
E	**BASE PLATFORM**	
	All of the base platform to be code	000
	except where marked f which will be	002
	g	001

Fig. 3   Manual simulation of possible plotter output.

particular unit of material is used with a particular "blob" shape its price is known and therefore as the plan is composed using these "blobs" the machine can not only compute the cost of the building but also produce a list of the materials and the amounts used. In this way as soon as a plan is completed an accurate cost can be obtained and any necessary amendments made before the output is committed to magnetic tape for subsequent reproduction for the contractors' use.

The finalised plan will be reproduced for contractual use, off-line on negative paper by a digital plotting device using the instructions now on magnetic tape but which were generated during the original creation of the plan on the 2250 screen. An example of the output thus achieved shown in Fig. 3 has been manually simulated.

Similarly the list of materials and their amounts which has been produced is used as input to a computerised bill of quantities program which the authority developed some time ago and which has previously relied on manually provided information but is now generated as the result of activity on the display unit.

These techniques involving the computer and in particular the 2250 have not endeavoured to cover 100% of any building project because to try and cover all contingencies and possible variations it is felt would not only be uneconomic but well nigh impossible and so the aim has been more in the region of 85% to 90% leaving the remainder to be dealt with manually. This remaining balance covers such things as special details designed specifically for a particular building, the method of constructing which is shown to the contractor by special drawings prepared for the particular purpose. The method of constructing the standard details covered by the "blobs" is shown to the contractor by selecting the appropriate detailed drawings from a library which having been once composed can be continuously drawn upon and expanded as development and research work goes on. The saving in what would normally be repetitious work is obviously great not only in manpower but also in time spent in the preparation of project drawings.

Conclusion

Just as a cost parameter has been attached to the "blob" shapes heat loss values which when totalled would give the heating requirements of the building and suggest alternative systems which would meet these requirements, their capital and running costs could also be attached. Further parameters will be that of weight and sound insulation. It is intended that the design of the units which infill the "blobs" should be carried out at the screen face and this further extension of the system together with site evaluation, and the more flamboyant perspective drawing programs are some of the developments which we shall be looking at in the very near future as well as sophisticating the techniques that I have already described.

To maintain high standards of design and construction in the face of the accelerating demand for buildings, is a challenge which is unquestionably there. The techniques which this paper describes provide the designer and the building team with a powerful tool with which to answer this challenge. However more tools with greater power will be required and it must be our endeavour to create them.

Biographical Note
P. E. Walter, A.R.I.B.A., has been engaged in Local Government service for some 22 years and is well experienced in the design and construction of local authority buildings using traditional "system" methods of construction. He is currently working as a member of a small team whose task is to produce a comprehensive system employing computer techniques, including man/machine conversations as an aid to design and management of building projects.

Graphical Output in a Research Establishment

F. M. LARKIN

U.K.A.E.A. Culham Laboratory

Abstract

This chapter discusses the use of graphical output facilities, particularly a programmable microfilm recorder, in an open-shop, scientific computing establishment. Many of the examples which are presented are drawn from the normal course of research at the U.K.A.E.A. Culham Laboratory, and illustrate the particular value of convenient graphical output to a research scientist.

Introduction

The distinguishing feature of a scientific research establishment is that its first objectives are knowledge and understanding. Although there may be a hope that the fruits of research will ultimately find commercial exploitation, usually the immediate motivation of an individual worker is to elucidate the properties of the object of his research, and to relate them to the existing body of knowledge.

It is not an accident that expressions commonly used to describe successful research lean heavily upon visual imagery. For example, we speak of "the light dawning", "a flash of insight" and "getting a fresh viewpoint". Indeed, unsuccessful research is often likened to "groping in the dark"! The human brain seems to be well adapted to handling pictorial concepts, and these are often used by scientists even when they are not strictly necessary. It is possible to design structures or solve problems in co-ordinate geometry without actually drawing pictures, but, especially in an unfamiliar situation, without even a rudimentary diagram most people would feel something lacking in their understanding of the problem. This is particularly true in research work which, by its very nature, deals with the unfamiliar situation in an attempt to become familiar with it.

In a scientific research establishment a computer is simply another research tool, to be classed with experimental equipment, mathematics, libraries, journals and symposia. The common feature of these "tools of the trade" is that they are all concerned with information—as sources or repositories, or as devices for transmitting or manipulating information. A computer, of course, is the information processing tool *par excellence*, and like any other tool it should, as well as being accurate and reliable, be responsive to the demands of its user. Oversimplifying somewhat, one might identify the role of a computer in research as a tool, complementing theoretical analysis, for exploring and sifting the consequences of hypotheses. However, the essential ingredient of the process is the delightful prospect of *unexpected* consequences, so that automatic presentation of the computational results in graphical form is vital to ensuring the responsiveness of the tool.

It is commonplace now that automatic graphical output is valuable in saving the drudgery of hand plotting, in order to obtain a broad view of the structure of the solution to any computing problem. Furthermore, though the development may not be cheaper than working by hand, it makes work practical that could not otherwise have been attempted in any reasonable time. However, the extra feature which marks it as essential in any research-oriented computing is its capacity for comprehensible display of novelty. Often, a result for which one is *not* looking may be more valuable than the result one wants. Horace Walpole coined the word "serendipity" to denote a faculty for making important discoveries by happy accident, and I think it is fair to describe automatic graphical output as a "serendipitous" technique.

In the examples presented in the Appendix (page 141), I illustrate the point with pictures of some recently discovered concepts; although in themselves these may not be of world-shattering importance, I think they are of more than parochial interest. All the examples share the properties that their essentials are easy to explain in pictorial terms and rather difficult in any other way; at the time they were quite unexpected, but having once seen them one feels they should have been discovered long ago. They were all discovered using automatic graphical computer output, and I venture to suggest that without it probably some would still be unknown and certainly ill-understood.

Computer Applications at the Culham Laboratory

In order to prepare the ground for the examples I would like briefly to explain something of the research program of the U.K.A.E.A. Culham Laboratory. The main objective of the Laboratory is to investigate the feasibility of producing useful power by means of controlled thermonuclear reactions. Roughly speaking, this means creating conditions in which the nuclei of a light element, deuterium or tritium, may be induced to combine in a controllable, self-sustaining fashion into helium nuclei, releasing energy in the process. In practice the only credible approach to this problem seems to be to use high intensity magnetic fields for confining the gas (since the temperature is great enough for it to be highly ionised), for a time long enough for nuclear fusion to take place.

Unfortunately, under such conditions the behaviour of the ionised gas, or "plasma" as it is called, is extremely complicated. As a result, the design of suitable magnetic confining fields, some of which are rather complex in shape, can be very critical.

The computational work resulting from this program is not untypical of any moderately large research or development project. Some one hundred scientists, who may be engineers, physicists or mathematicians, use the computing facilities on an open-shop basis.

Most computing problems originate either from individuals or from small teams, who regard computation as part of their own job; they will either do their own programming or take a close interest in that work when it is done for them by the computing section. In nearly all cases, problem originators will run their own programs as soon as these are operational, and will do any development which seems necessary in the light of the initial results. Experience suggests that this method of working is well suited to the research environment; many scientists like to be as self-sufficient as possible since, when grappling with a research problem, it is difficult to separate the intellectual processes involved from the associated, relatively mundane tasks like computing and graph-plotting.

Nearly all computing jobs could be described as "one-off", rather than "production", and many of these are run with the objective of getting a feeling for the general character of a problem, rather than to obtain specific numerical results. There is, however, a wide variety in the complexity and running times of programs. Most users employ the local graphical output system, GHOST (Larkin 1968), as a matter of course, often using printed output only for reference purposes, or when high accuracy is required. A typical day's work will result in some 200 computing jobs and 400 graphs, with a few 35 mm. slides and the occasional scene of ciné film.

Equipment

In this situation rapid turn-around, both of printed and graphical output, is of paramount importance. Figure 1 illustrates schematically the particular hardware compromise adopted at the Culham Laboratory in an attempt to provide for the requirements of the various users. The arrows indicate the directions of information flow.

A sequence of so-called "background" jobs enters the main KDF9 computer from the card reader, via a magnetic tape buffer. Printed output is accumulated in the magnetic tape output buffer and fed to the line printer at the discretion of the operators. Simultaneously, a number of teletype users may be modifying information in private disc files, interrupting operation of the background job whenever any disc accesses are required. These "foreground" users can also suspend the background stream in order to enter short jobs for almost immediate processing. Both background jobs and jobs initiated by foreground users can cause graphical output to be sent either to the magnetic tape graphical output buffer, or to the on-line CRT display. Normally, graphical output from a number of jobs is accumulated on the tape buffer, awaiting the next processing run.

At the discretion of the computer operators, the stored graphical output undergoes a further processing phase which produces a control tape for the Benson–Lehner Model 120 Printer/Plotter. This tape is then transferred to the off-line B–L 120 where it may be used to produce pictures many times over if required. Once the control tape

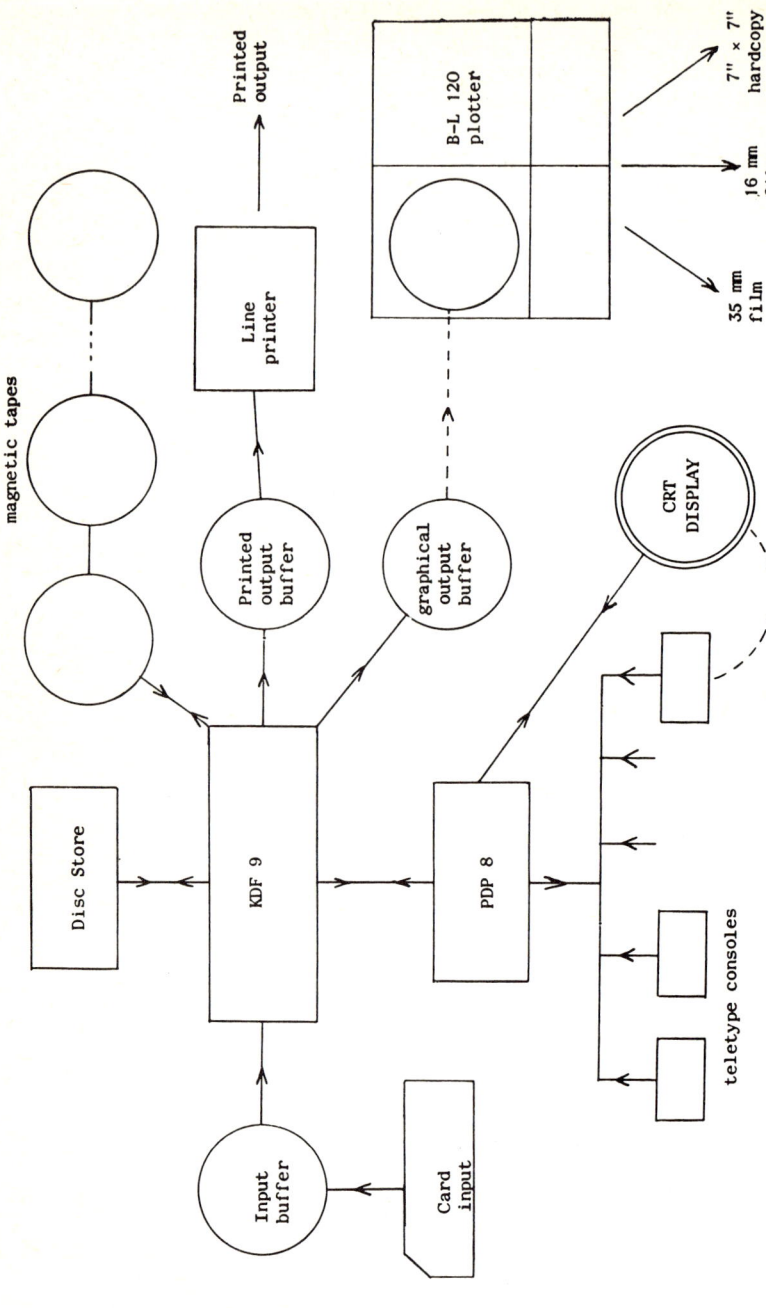

Fig. 1 Culham Laboratory hardware layout, with arrows showing direction of information flow

is mounted and set in motion, operation of the plotter is automatic, very little operator attention being necessary.

Use of the system varies, of course, with a user's experience and current requirements. However, it is quite feasible to take, say, half a dozen short runs, using the C.R.T. display and an on-line teletype, to debug and develop a new program, and on the same day to get permanent graphs of production results from the B–L 120.

General Remarks

It is fair to say that *convenient* graphical output facilities form an indispensable part of computing in a scientific research establishment. I emphasise the word "convenient" here since, although very similar hardware facilities have been available to Culham Laboratory for some time, it was not until the GHOST software was introduced that the majority of computer users began seriously to exploit the possibilities of graphical output.

The largest single areas of application are the automatic plotting of simple output functions $f(x)$ against x, say, which one might term "one-dimensional" graphs, and the display of experimental results. In a sense this use of graphical output is simply as a timesaver. However, it is unlikely that many two-dimensional displays, such as contours, for example, would be attempted if hand plotting were the only method. Three dimensional displays, in the form either of stereo pairs of ciné film, would be out of the question if automatic plotting were not available.

To sum up then, *convenient* computing and graphical output facilities, especially of an on-line, interactive nature, provide the scientist with very powerful research tools. We are still learning how best to exploit these tools, and the companion facility of graphical input, but experience to date is very encouraging. Pictures possess an impact, or immediacy, which seems different in character from numerical knowledge and makes graphical output peculiarly useful in research.

5. *Acknowledgements*

The author would like to express appreciation to numerous colleagues at the U.K.A.E.A. Culham Laboratory for help and advice freely given, in particular to Dr. K. V. Roberts and Dr. A. Gibson.

REFERENCES

Gibson, A., and Taylor, J. B. "Single particle motion in toroidal stellarator fields", *Physics of Fluids*, **10**, 8, 1967, pp. 2653–2659.

Larkin, F. M. "A combined graphical and iterative approach to the problem of finding zeros of functions in the complex plane", *Computer J.*, **7**, 3, October 1964, pp. 212–219.

Larkin, F. M. "A single loop conductor for producing a magnetic well", CLM-R37, 1964.

Larkin, F. M. "The structure and implementation of GHOST." *Computer Bulletin*, Vol. 12, No. 8, Dec, 1968, pp. 286–291.
Roberts, K. V., and Berk, H. L. "Non-linear evolution of a two-stream instability", *Phys. Rev. Letters*, **19**, 6, 1967, pp. 297–300.

Biographical Note

F. M. Larkin graduated with B.Sc.(Hons. Maths.) from Imperial College, London. He worked for four years in the field of Nuclear Engineering at A.E.R.E. Harwell and Rolls Royce. He is now working as Numerical Analyst at U.K.A.E.A. Culham Laboratory for Plasma Physics and Nuclear Fusion Research.

Appendix to
Graphical Output in a Research Establishment

F. M. LARKIN

The following miscellaneous examples, though naturally of a very technical nature, nevertheless demonstrate conclusions which are easily grasped and whose value was mentioned in the body of the chapter.

(i) Magnetic Well

One type of magnetic field which has received attention as a possible configuration for the containmnent of thermonuclear plasma is the so-called "magnetic well". Physically, this is simply a localised region in space containing a relative minimum (not a zero) in magnetic field strength. Magnetic wells come in a variety of shapes and sizes, but all are fundamentally restricted by the fact that the magnetic field B must satisfy the relation

$$\nabla . B = 0$$

which among other things makes it necessary for any true magnetic well to be a strictly three dimensional object. It is this essentially three dimensional nature which makes pictorial representation so useful for the understanding of magnetic wells, and especially of their relationship to the conductor configurations which produce them.

At school we learn about the magnetic fields produced by an infinite straight wire and a circular loop, but I would like to suggest that familiarity with these simple configurations is a feat of visual memory, rather than of intellect or imagination! Let me ask you to imagine a filamentary, electrical conductor shaped like the seam of a tennis ball; that is not a trivial mental exercise, but it should not be too difficult since everyone has handled a tennis ball and inspected the shape of its seam. However, now let me ask you to visualise the shape of the magnetic field associated with a current flowing in the conductor—that is not quite so easy! Figure 1, actually one member of a stereo pair, illustrates the situation, which is that of the simplest method for producing the simplest type of magnetic well (Larkin, 1964). One very quickly becomes familiar with conductor-field configurations of this sort, even to the extent that one tends to forget how obscure they were before the picture was available; in other words one remembers the picture.

(ii) Orbits of Charged Particles

One of the most fundamental problems at the Culham Laboratory is that of determining the orbits of charged particles in a magnetic

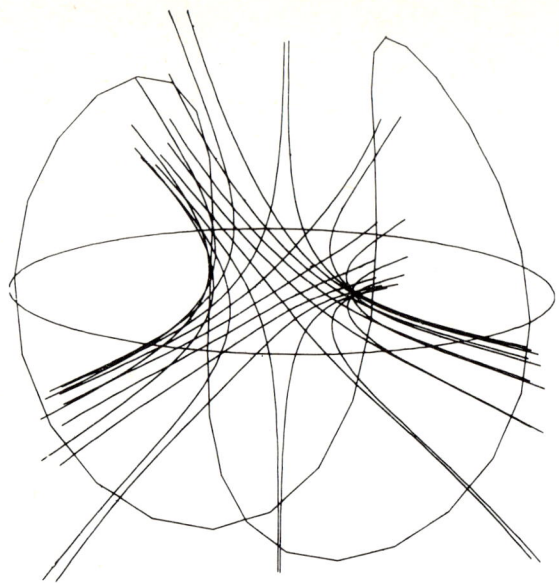

Fig. 1 Magnetic field lines due to tennis-ball-seam conductor

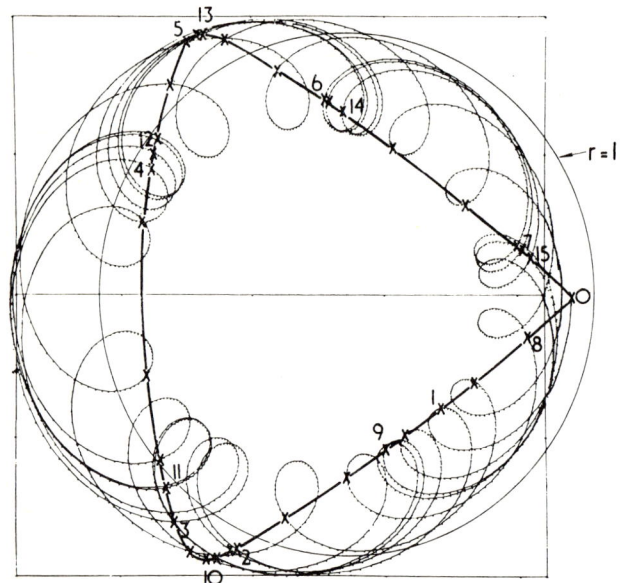

Fig. 2 Passing particle confined

field. For a given design of plasma containment apparatus it is important to know which types of particle are likely to be confined, and how they will be distributed. This is an ideal subject, both for computation and for graphical display. The instantaneous behaviour of a charged particle in a given magnetic field may be regarded as subject to known, classical laws, and the complete orbit, or at least a portion of it, may be found by integrating the equations of motion as an initial value problem.

Figure 2 shows a minor cross-section through a toroidal plasma containment field. The curling line illustrates an azimuthal projection of the track of a single charged particle moving in the field. Although the motion is fairly complex, its roughly cycloidal nature was not unexpected. However, Figure 3 illustrates the result of slight alteration of the initial conditions of the particle. This shows that after escaping from an initial, cycloidal type of confinement the particle is trapped in an outer region, where it progresses by means of a sequence of reflections. When this latter behaviour was first observed (by means of computer and graphical output, *not* experimentally) it was quite unexpected. As a result, detailed investigations of various particle trapping modes were carried out (Gibson and Taylor, 1967) and some valuable results were obtained. Figure 4 shows an orbit of a different type.

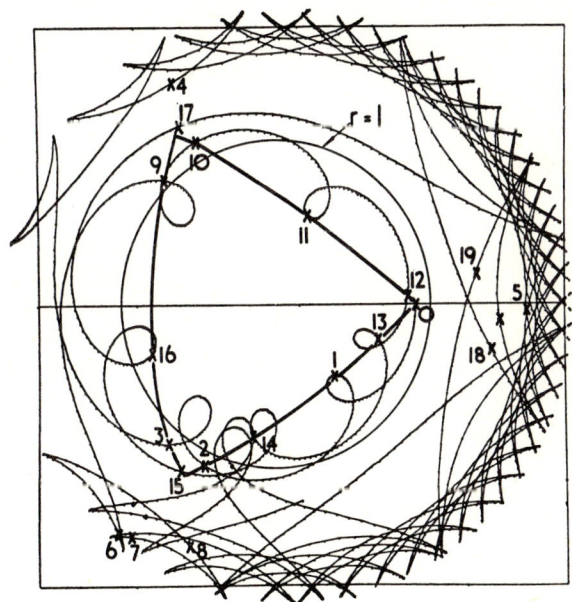

Fig. 3 Passing particle escapes

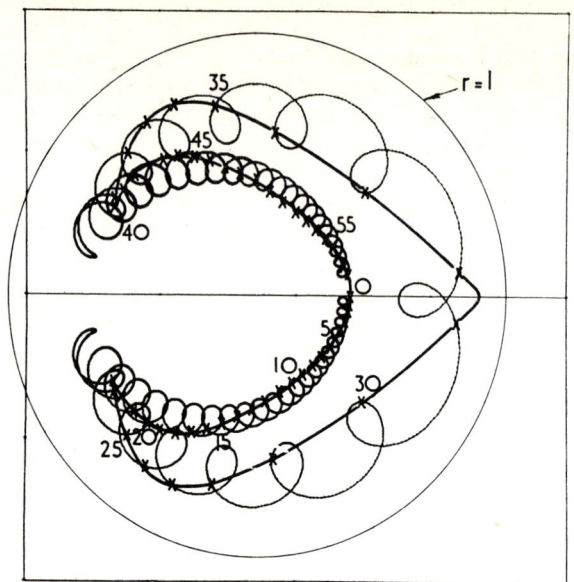

Fig. 4 Blocked particle confined

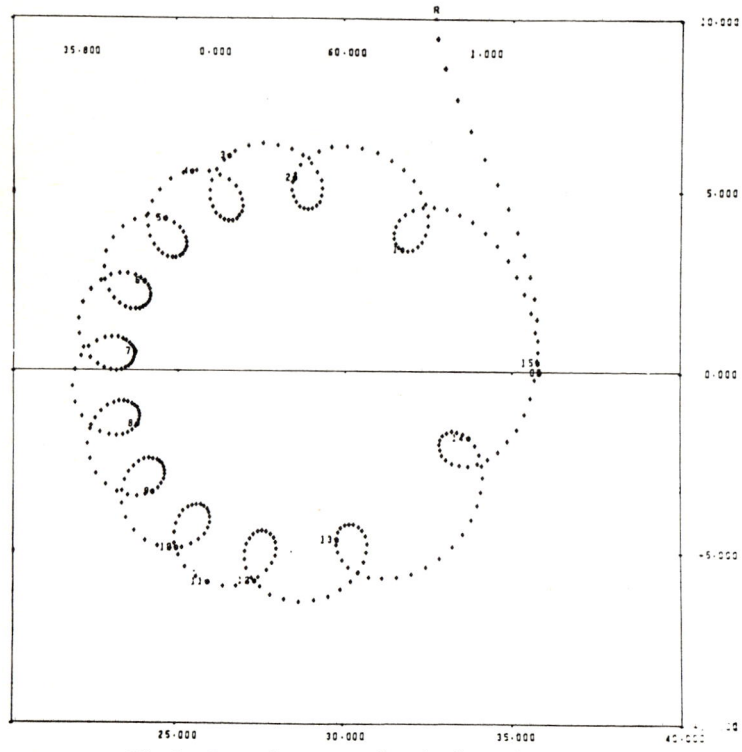

Fig. 5 Correctly computed path of escaping particle

(iii) Detection of Computer Fault

As an example of an unexpected bonus from the use of graphical output I would like to mention an incident that arose in connection with particle orbit studies. In March of this year, Dr. Gibson was using a well proven program for design calculations on a toroidal containment field. While investigating the trajectories of marginally confined particles he observed one orbit which should have looked like that shown in Figure 5, but in fact looked like that shown in Figure 6. Apparently, the particle had received some severe jolts at a part of its orbit where such a thing was physically impossible, although most of its trajectory looked fairly reasonable. In the circumstances, Dr. Gibson immediately informed the computer manager that the KDF9 was suffering from a small, intermittent fault in the arithmetic unit—the computer user's nightmare! It took two weeks for the engineers to find the fault and correct it.

(iv) Behaviour of "Phase Fluid"

I would like now to cite an example* of the discovery, through the medium of computation and graphical output, of some quite new

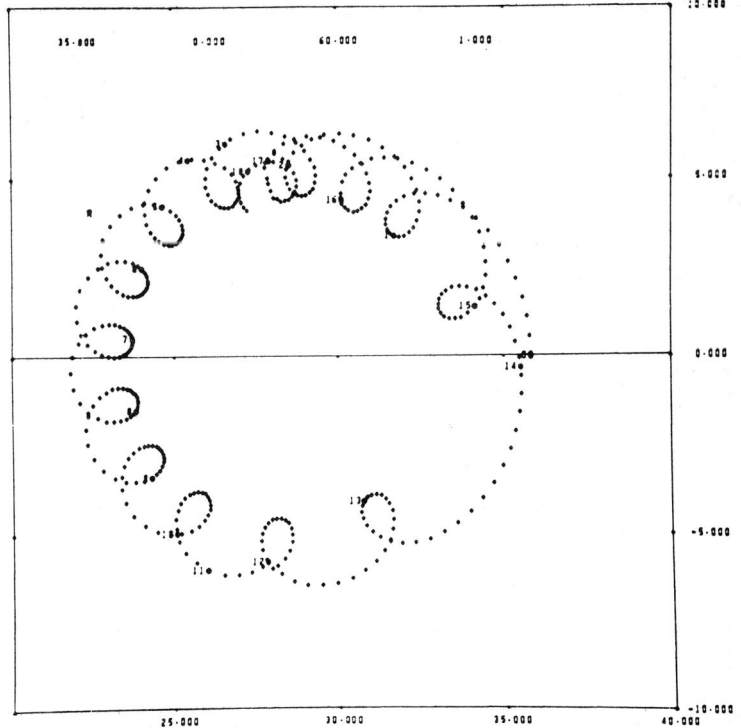

Fig. 6   Spurious irregularities in particle path

* Work done at the University of California, San Diego.

physical phenomena. It is known that a plasma is unstable if the electron velocity distribution function is bimodal, but only the initial linear phase of this so-called 'two-stream' instability has hitherto been amenable to theoretical analysis. In a study of the later, non-linear phase of the two stream instability, Roberts and Berk (1967) have computed the time development of a simplified model.

The model is described by the usual one dimensional Vlasov–Poisson equations

$$\frac{\partial f}{\partial t} + v \cdot \frac{\partial f}{\partial x} - \frac{\partial \phi}{\partial x} \cdot \frac{\partial f}{\partial v} = 0$$

$$\frac{\partial^2 \phi}{\partial x^2} = \omega_p^2 \left[1 - \int f \frac{dv}{v_0} \right]$$

where $f(x, v, t)$ represents electron density in phase space, ϕ is proportional to the electric potential and ω_p and v_0 are constants. Roberts and Berk chose simple initial conditions in which f takes either of the two values 0 or 1 in prescribed regions of phase space. The subsequent behaviour of the system is then reminiscent of the motion of

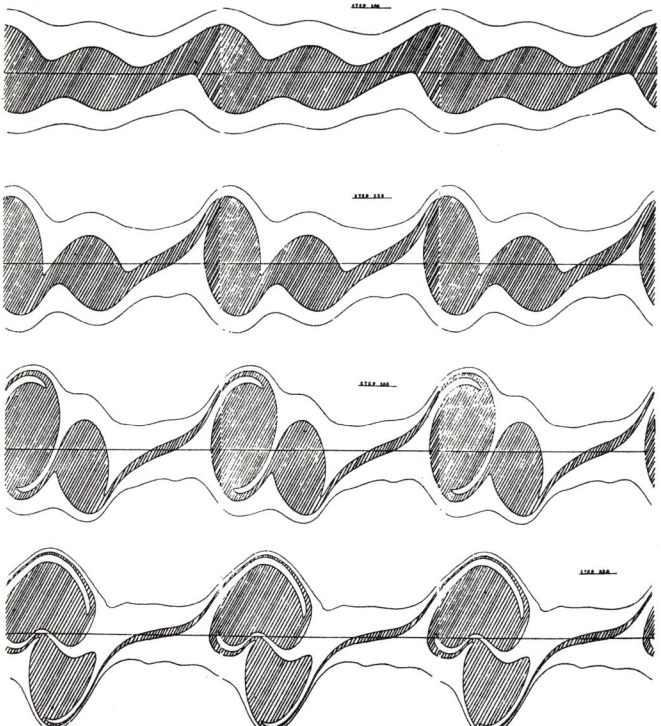

Fig. 7 Formation of holes in "phase fluid"

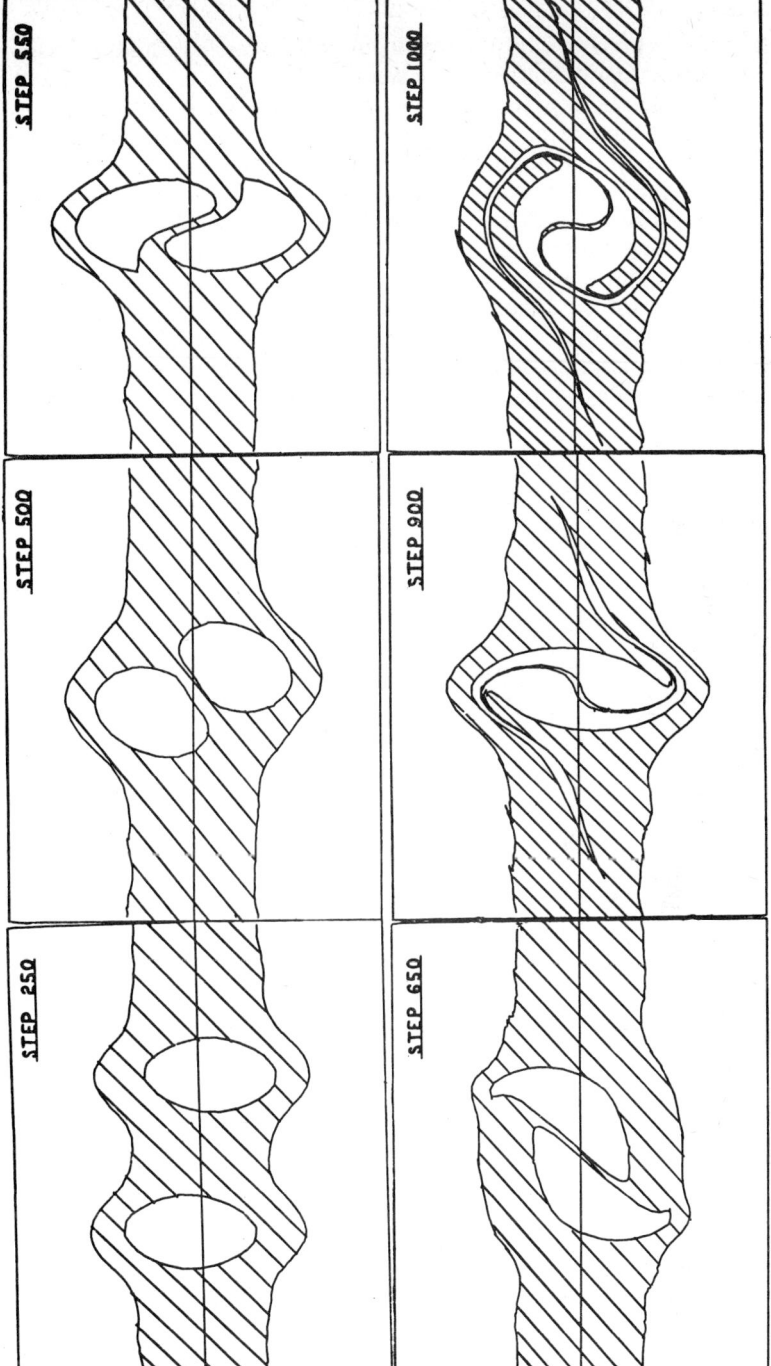

Fig. 8 Interaction of holes in "phase fluid"

interacting globules of a homogeneous incompressible fluid, such as water, suggesting the term "phase fluid".

Figure 7 shows several successive stages during the evolution of this "phase fluid" model of a two stream instability. A significant feature is the formation of holes in the fluid which persist and appear to interact, behaving as if they were individual entities which attract one another and fuse together. Figure 8 illustrates the results of a slightly different calculation which exhibits similar phenomena. The results of these calculations, which would have been almost incomprehensible without the use of graphical display, have suggested new lines of research in such diverse fields as fluid dynamics, meteorology, statistical mechanics, plasma physics, theoretical astronomy and particle accelerator theory—quite an achievement for a computer and graph plotter!

High Energy Physics Applications

P. M. BLACKALL, J. C. LASSALLE,
C. VANDONI and A. YULE

CERN, Geneva, Switzerland

Abstract

Types of problems in High Energy Physics which can be tackled using graphics—suitability of existing hardware and software—goals and experiences at CERN.

Introduction

The aim of high energy physics research is to understand the composition of matter by studying the properties of sub-nuclear particles and by determining the forces that govern their interaction. The experimental techniques used involve directing a beam of high energy particles of known kind and energy onto a target of known nuclei and observing the secondary particles produced within the target volume. A typical experiment requires the recording, measurement and analysis of thousands or hundreds of thousands of such nuclear interactions or events.

During the last ten years there has been an increasing use of digital computers in high energy physics. Computers have become an essential tool in the planning, data acquisition and data analysis phases of experiments as well as in the control of particle accelerators, beam switchyards and particle detectors.

With the use of digital computers there has followed a need for graphics for the rapid assimilation of processed data. Graphics is being used in the passive, active and interactive forms. In the passive form graphic output is generated with mechanical plotters and microfilm recorders; in the active form character displays and point storage tubes are used for monitoring processes and in the interactive form both graphic input and output are generated using graphic displays equipped with keyboard and lightpen sensing devices.

Interactive Graphics at CERN

A description of the interactive graphics work at CERN can illustrate the use of graphics in high energy physics. The work has been carried out on a CDC 250/3398 buffered display system connected to a CDC 3100 computer with 16 K words of core memory and 2 M words of disk storage.

Picture by Photo CERN

Fig. 1 A stereo photograph of particle trajectories and interactions taken on the CERN 2 metre hydrogen chamber

Generation and Measurement of Artificial Bubble Chamber Events

A bubble chamber is a particle detector which enables a photographic record to be made of particle trajectories, and of particle interactions occurring from the collision of beam particles with nuclei of the chamber liquid. Images of the particle trajectories are recorded, usually, on three stereoscopic photographs (see Figure 1). The chamber is operated on a fixed cycle recording all events which occur in its volume as it cannot be triggered to operate only when interesting events occur.

In the first phase of the analysis of the recorded events, the triads of photographs are visually examined to select those triads containing events of interest. Measurements of points on the track images of the interesting events are then made with either manual or semi-automatic measuring devices. The events are reconstructed in space from the track measurements made on each stereoscopic view together with geometric and optical data of the bubble chamber. This phase of the analysis, geometrical reconstruction, is carried out by standard programs applicable to any class of event or to any bubble chamber.

In the design of a new bubble chamber it is important to consider the problems of data analysis. With the proposed camera positions will a physicist be able to examine the triads of photographs and recognise interesting events? Will the standard geometrical reconstruction programs cope with measurements of events from the chamber? Answers to questions of this kind can be rapidly obtained

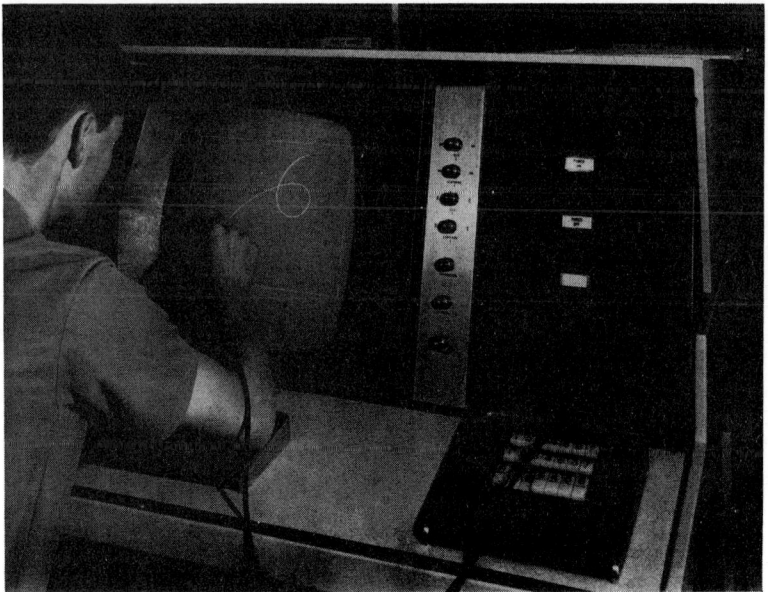

Fig. 2   Physicist using a lightpen to "measure" points on a track image of an artificial event for a new bubble chamber

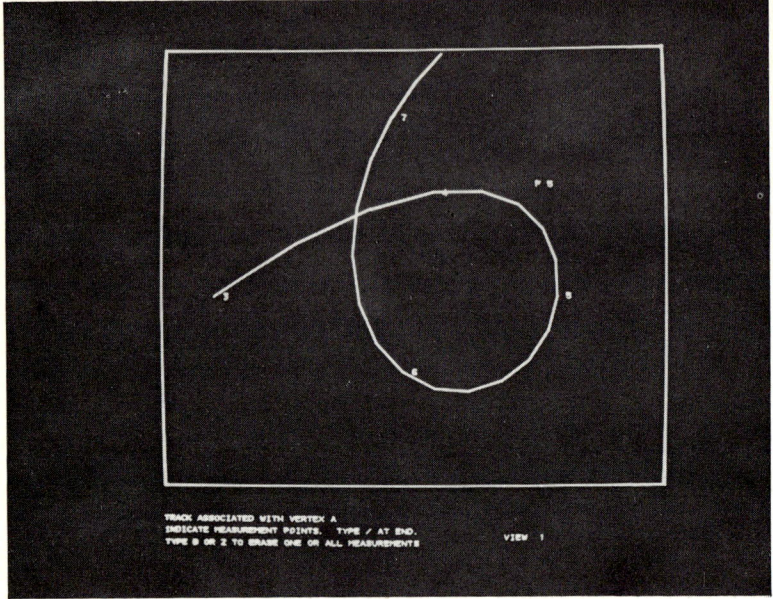

Fig. 3 Magnification of the track image of Fig. 2

with the aid of interactive graphics. Figure 2 shows a physicist using a graphic display to make "measurements" with the aid of a lightpen on a track image of an artificial event for a new chamber. The track image had been previously calculated from numeric physical data entered at the keyboard. Figure 3 shows a magnification of the same track image permitting the physicist to make several measurements along a short portion of the track.

Analysis of Spark Chamber Events rejected by Automatic Measurement Systems

The spark chamber is a particle detector which, like the bubble chamber, can be used to visualize events by recording images of particle trajectories usually on two stereo photographs. Although the spark chamber cannot provide the same precision as a bubble chamber in the location of a particle trajectory, it has the advantage of being able to be triggered on the occurrence of an interesting event. Thus the photographs obtained from spark chambers contain a high proportion of events and each photograph contains only the particle trajectories of a single event together with possibly one or two old incident beam tracks.

The particle track information on film is converted from analogue to digital form by a special-purpose flying spot digitiser connected on-line to a computer. A pattern recognition program extracts from the raw digitised data those measurements corresponding to particle tracks and deduces whether the particle tracks form an interesting event. Depending on the result of the analysis, either event data is written onto magnetic tape for subsequent input to a geometry program or rejected event data is written onto another magnetic tape for subsequent examination. The rejected events require further visual examination to determine whether there was no event or whether the pattern recognition program was treating this class of topology incorrectly. In the latter case corrective action can be taken at the display by a trained operator and the required event data written onto magnetic tape. Although the rejected events represent only a few per cent of the total number of events, it is important to understand the reasons for their rejection otherwise a bias could be introduced into the experimental results.

Figure 4 shows a display of the spark positions in a stereo view of an event recorded from a spark chamber. The pattern recognition program could not resolve in one stereo view an ambiguous situation in the pairing of tracks from two apices.

Figure 5 shows the stereo view redisplayed with the track pairing resolved after corrective action by an operator. Figure 6 shows a display of both stereo views enabling the operator to check whether the manual track pairing in one view agreed with the automatic track pairing in the other view.

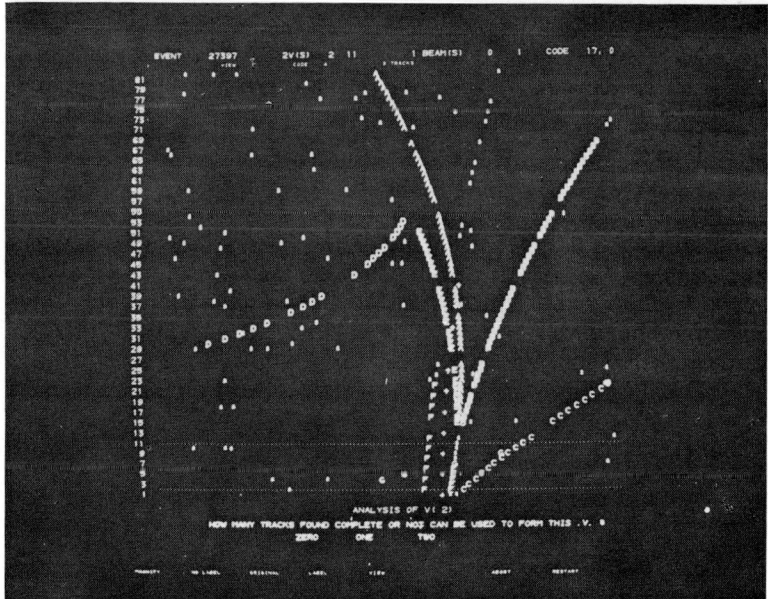

Fig. 4 Display of a stereo view of a spark chamber event rejected by an automatic measuring system

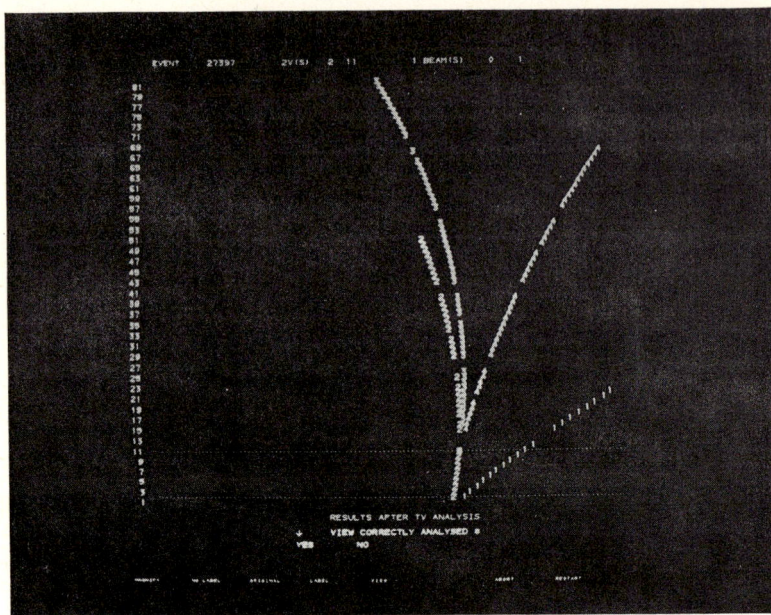

Fig. 5 Display of the stereo view in Fig. 4 after corrective action at the display by an operator

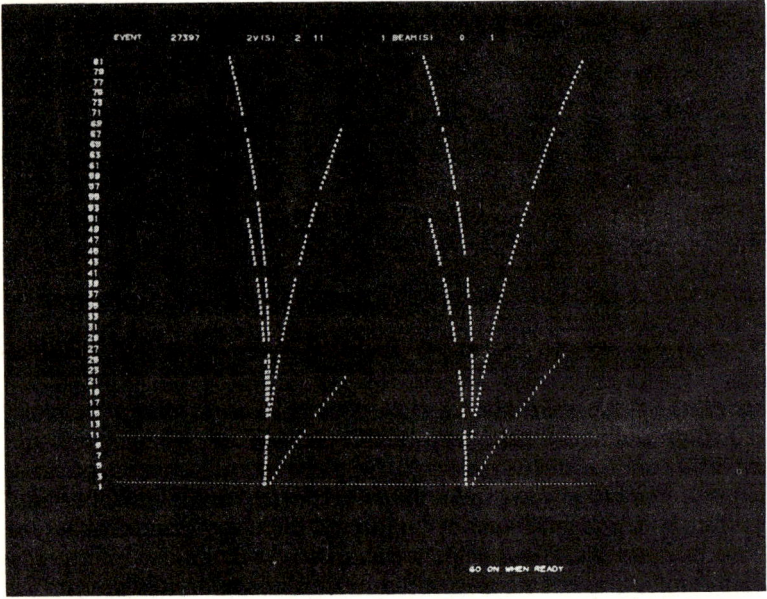

Fig. 6 Display of both stereo views showing track pairings selected by the operator in one view and track pairings selected by an automatic measuring system in the other view

On-line Mathematical Analysis

When considering the use of computers in high energy physics the emphasis is usually centered on the requirements of the experimental physicist. However, the theoretical physicist can make good use of interactive graphic facilities as an aid in mathematical analysis. An interesting development in this direction is the Culler-Fried On-line Console System.[1] This system permits a user to carry out numerical calculations from a keyboard using the standard mathematical operators. A calculation is performed by pressing a series of keys corresponding to operator and operand actions followed by a display key action to display the results of the calculation. In its simplest mode of operation the system provides an expensive desk calculator whereas in its common mode of operation the system provides a powerful function calculator. Functions are approximated by vectors of 100 or more numeric values and the basic mathematical operators operate on the complete vector of values.

To obtain some user experience of this type of interactive system an experimental single console system has been implemented at CERN. The system provides identical facilities to those of Culler using the typewriter keyboard of a CDC 250 display. Figure 7 shows the display of a complex function in which the real part is plotted as abscissa and the complex part as ordinate. The function is of the form:

$$F = (2l + 1)[\delta_{l0} - e^{-i\pi\alpha_0} i^l e^{i2q^2\pi\alpha_1} j_l(-2q^2\pi\alpha_1)]$$

where l, α_0 and α_1 are parameters and q^2 is the independent variable.

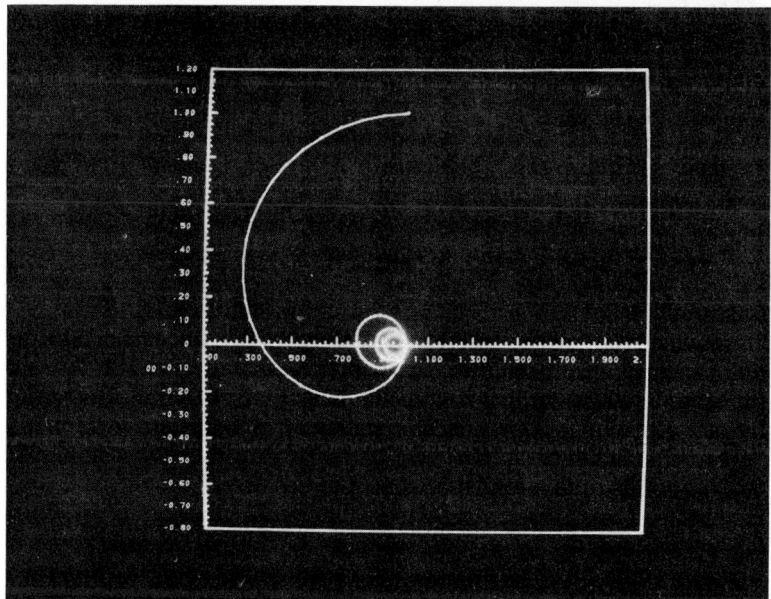

Fig. 7 Display of a complex function generated with the "function calculator"

```
USER LEVEL = I      KEY= M
LIST LOAD @ * 6. 283   * Y STORE @ LOAD X * 3
. 1415926   CONJ + @ STORE A COS STORE R LOAD
A SIN STORE I LOAD @ CONJ STORE @ SIN / @ S
TORE B * R STORE X LOAD 1  - X STORE X LOAD
B * I CONJ STORE Y LOAD @ COS CONJ + B / @ S
TORE T * I STORE U LOAD T * R CONJ STORE V L
OAD T * 3 / @ - B STORE S * R STORE W LOAD
S * I STORE Z LIST
```

Fig. 8   Program of key actions to generate the complex function

The behaviour of this function for various values of the parameters and ranges of the independent variable was examined in the interpretation of an apparent resonance in a cross-section.

Figure 8 shows the program of key actions required to compute the function. The system provides a facility for storing user programs which may subsequently be executed by a single key action.

Potential Applications of Graphics

Graphics is being used or is planned to be used in almost all phases of experiments. Many of the applications require an active display with which a physicist may monitor and modify the particular process under observation. The areas of application of active graphics are control and monitoring of particle accelerators, beam switchyards, particle detectors, data acquisitions systems and data reduction systems.

In addition to the applications given in section 2 potential applications of interactive graphics are beam optics design, magnet design and statistical analysis of experimental data. The successful exploitation of on-line statistical analysis of experimental data would probably require multi-console facilities to meet the needs of physicists. The geometric and kinematic information for all the events in an experiment is collected on one or more magnetic tapes and is subjected to statistical analysis. Aggregates of the data in the form of histograms and scatter diagrams are generated to show correlation between computed quantities and to reveal biases or systematic errors in the data. Frequent use is made of these statistical analysis programs with different selection criteria and different computed quantities. At present the statistical analyses generate large quantities of histogram or scatter diagram plots on line printers using off-line batch processing systems. Development projects[2,3] are in progress to produce on-line systems for this type of statistical analysis. However the provision of a general interactive graphics facility with access to the computing power of a large computer is a preferable development.

Requirements for Hardware and Software

The three types of display commonly used in high energy physics applications are the character display, point storage tube and the interactive graphic display. Ignoring the question of cost, approximations to a suitable character display are available from most manufacturers. The introduction of the larger Tektronix storage tube (type 611) closely satisfies the primary requirements of resolution and storage for a cheap graphic display. The specifications of the additional facilities developed for this tube by Computer Displays Inc.[4] suggest that a complete inexpensive graphic display will soon be available.

Approximations to a suitable interactive graphic display are also available from manufacturers. However the hardware requirements are not limited to those of the display and its associated buffer controller or small computer. The applications in high energy physics require access to the computing power of a large computer (e.g. CDC 6600, IBM 360/75, UNIVAC 1108). It is thus desirable that the display is attached to its own small computer for the purpose of refreshing the display image and for dealing with any local interaction i.e. keyboard input or lightpen action. The large computer is then interrupted only for computational purposes.

The allocation of separate functions to the display computer and to the large computer requires appropriate extensions to the operating system of the large computer. It is not economic to permanently assign part of the core memory of a large computer to an interactive graphics application requiring frequent manual intervention. The display computer can deal with all local interaction, e.g. storing a series of character or lightpen inputs—drawing with a tracking-cross. The large computer need only be interrupted to perform a memory swap to bring in the application program from a rapid access backing store, to compute and to perform a further memory swap to restore the original background program.

Programs for high energy physics applications are generally written in FORTRAN by the physicists concerned with the application. It is therefore necessary to provide graphic facilities as extensions to FORTRAN in the form of application subroutines. Subroutines are required to map the users data co-ordinate system into the display coordinate system, to display points, vectors, arcs, axes, labels and text, to modify the basic display conditions (i.e. character size etc.) and to input and scale data from the keyboard and lightpen. In general the user requires to retain the data structure of his FORTRAN program although an ability to assign identification and structure to display items and to construct subpictures is desirable.

The software of the display computer should provide facilities for performing a series of identical input operations, e.g. character input—lightpen hits—lightpen tracking, with local editing facilities to delete either the last operation or all operations.

The development of an interactive graphics application program requires the frequent correction and executing of the application program. Most installations operating large computers do not however have time-sharing systems. It would appear that until such systems are available the development of application programs will use the standard batch input facilities rather than the graphic display.

Conclusions

Active graphics is already used in most phases of a high energy physics experiment. With improvements in display hardware and a reduction in cost, we can expect an increasing application of active graphics in the monitoring and control of processes.

Interactive graphics is restricted by the lack of economic computing power. Many of the existing and potential applications require the use of a computer with a large memory and a fast central processing unit. Further effort is required on the part of manufacturers to provide an economic method for the use of single and multi-display systems with large computers.

REFERENCES

1. Culler, G. J. *User's Manual for an on-line system. On-line Computing*, pp. 303–324. Ed. Karplus, W. J., McGraw-Hill, 1967.
2. Ophir, D., Rankowitz, S., Shepherd, B. J., and Spinrad, R. J. *Multi-console computer display system.* Brookhaven National Laboratory, Report 11589, 1967.
3. McGee, W. C., and Howry, S. K. *On-line SUMX.* Report No. 320–3230, January 1968, IBM Palo Alto Scientific Center.
4. *Electronics,* pp. 50–51, February 19, 1968.

Biographical Notes

P. M. Blackall received the M.Sc. degree from Imperial College, London, in 1956. After working as Scientific Officer in the Mathematical Physics Division of the Atomic Weapons Research Establishment at Aldermaston and then as Systems Programmer in the Computing Department of the Central Electricity Generating Board in London, he joined C.E.R.N. (Data Handling Division). He is at present responsible for development projects in remote access, graphics and on-line computing.

J. C. Lassalle, Ingénieur Institut Polytechnique de Grenoble (Electro-technique 1961—Mathematiques Appliquées 1961). In 1964 he joined CERN (Data Handling Division), Geneva, Switzerland, and is engaged in the development of programs for automatic analysis of spark chamber pictures.

C. Vandoni, B.S. (University of Milan). From 1959–1964 he worked as Systems Programmer in the Electronic Division of Olivetti in Milan. He joined CERN (Data Handling Division) in 1965, and is at present responsible for implementing the graphical aided console system for on-line mathematical analysis.

A. P. Yule, Dip.Tech. (Applied Physics). From 1964–1967 he was Research Assistant in the Computing, Cybernetics and Management Department of Brighton College of Technology. He joined CERN (Data Handling Division) in 1967, and is working on the production of software for a graphic display and the simulation of particle tracks in bubble chambers.

Mechanical Design Using Graphics

B. T. TORSON
Rolls Royce Ltd

Introduction

The first part of this paper will describe the limits in terms of both equipment and applications. This will be followed by a brief description of the technical application of computers at Rolls Royce which led up to the work on computer graphics. The main part of the paper is split into three sections describing past, present and future applications in Mechanical Design at Rolls Royce.

Limitations

The type of equipment whose application will be described is interactive computer graphic. That is it can be used to display alphanumeric characters, curves and shapes calculated by a computer under the control of the user, or allows the input of these things by means of such items as alphanumeric keyboard, button box and light pen. The distinction will be made between the more sophisticated kind which has a light pen and allows the user to "sketch" on the screen and the simpler kind which can display geometric shapes but which allows action by the user only via a keyboard.

The paper will be concerned solely with applications of interactive computer graphics originating in the design and analysis of aero-engines; it will not be concerned with any of the other activities which take place at Rolls Royce such as manufacturing aero engines or designing and manufacturing cars and nuclear reactors. The most important characteristic which aero-engine parts possess which is relevant to this paper is that they are designed to fulfil functional requirements as closely as possible, and aesthetic considerations play no part at all.

Technical Application of Computers at Rolls Royce

The first digital computer was installed at Rolls Royce in January 1956 to carry out calculations for several of the technical departments. Until a few years ago, although the growth in the technical use of computers was large, the users were still the staff of the technical departments, that is performance engineers, aerodynamicists, stress engineers etc. The designer, the man who worked on a drawing board to produce design schemes using the information, advice and instruction of the technical specialists, did not use the computer himself. This is in distinction to former times when there were no technical specialists and designers did their own calculations.

The designers started to use the computer themselves on certain problems and had several general comments to make; among them

(i) input sheets for entering data (via punched cards) into a computer program was not the ideal way; it was too susceptible to error.
(ii) output in tabular form was far from ideal.
(iii) the elapsed time between problem formulation and solution, consisting of several separate computer runs was too long.

About three years ago a survey with technical users of computing on their present and future requirements took place. As part of this survey, discussions involving many designers were held to establish their requirements. One of the constraints that had to be taken into account was that over the years a very large number of application programs had been written (almost entirely in FORTRAN), and designers could want access to these programs. From all this and other work emerged the conclusion that Rolls Royce should obtain an interactive computer graphic device on rental in order to evaluate its application in the design, development and manufacture of aero-engines. This was done and in July 1967 an IBM 2250 Model I was installed attached to a 512K byte System 360 Model 65 with direct access and magnetic tape storage facilities. Subsequently this has been changed to a 2250 Model III.

The conclusion that graphics had a potential and that it ought to be evaluated arises from the fact that the firm is in business to make something to sell at a reasonable profit, and the cost, delivery and quality are three of the most important factors. If any of these three can be improved, then the firm is in a better competitive position.

With graphics, it was considered possible to improve in all three of these areas, though in quite what proportion, a fairly senior manager would have to decide. For instance, taking the same time for a job, but with quicker turnaround from the computer, should allow a much more thorough job. Conversely, maintaining the same performance, it would be possible to finish it considerably earlier.

Phase I

After discussions with the Chief Designer and Senior Design Managers it was decided that

(i) no demonstration programs would be written; from the outset the intention would be to implement graphics programs which would be of significant value to the Design area.
(ii) No new "linked" programs would be implemented, i.e. each program would do a single job, instead of being one of a suite of programs in theory capable of very large tasks.
(iii) No new methods of calculation would be attempted. Initial applications would be based on existing, programmed methods.

This allowed effort to be concentrated on the interactive graphics aspect and not spent on iteration convergence and similar problems.
(iv) It was necessary to have programs that were easy to use because it was considered undesirable to create specialist users and there were therefore very many potential users of each program.
(v) The design area, having chosen which programs should be implemented, and also bearing (iv) in mind, would also supply manpower to specify in complete detail the console procedures and facilities required from these programs.

This last decision was probably the most important single decision taken; it ensured that a potential user had to think out, and document, how *he* would like to take advantage of the new facilities. A consequence of this was that a sequence like

(*a*) input data,
(*b*) calculation exactly as in batch program,
(*c*) examine selected results,
(*d*) modify data and repeat from (*b*) until satisfactory results are obtained,

was quickly ruled out. What had been a single closed calculation in the batch mode became an almost infinitely variable combination of smaller calculations because of the ability to examine partial results and to have calculations performed in almost any order. In addition to this, there were several hundred potential users, who knew very little about computing, and two-day courses are being run, for about twenty designers at a time, to take them through the things they can do on a computer (as well as what the computer staff can do); they meet the people running the system, quite an important point; and they see the equipment itself, which is a matter of considerable interest to engineers.

Applications

The two applications now described are chosen to illustrate two additional advantages which were derived from the work. The first is a disc-design program. A disc here refers to an axisymmetric body rather like a gramophone record but with a thickness which varies with radius. Given material properties including allowed stresses and certain geometric information including limits, a conventional program exists which will compute the minimum weight shape of the disc. However the designer may wish to vary the maximum thickness or other restrictions to see the effect on weight, and the program contains several iterative procedures which cannot economically be made 100% convergent. Consequently several separate runs may be necessary taking a week or more.

The graphics version allows the designer to enter input parameters, the names and entered values being displayed on the screen; default values are displayed where appropriate. During this phase only the character display and alphanumeric keyboard are required. When the designer indicates that the input is complete the iterative procedure begins. The progress of the iteration is displayed and can be interrupted at any time and various options are available including return to input phase, modify iteration parameters and accept convergence. On convergence the computed shape and key thicknesses are displayed. The options available at this point include

(i) displaying plots of selected stresses or growths against radius,
(ii) obtaining permanent printed and/or plotted results,
(iii) Storing the input data in either a permanent or temporary file, the difference between the two files being that the contents of the temporary file are lost at the conclusion of the 2250 session.
(iv) returning to the input phase.
(v) selecting a different problem or at least a different version from one of the files.

On return to the input phase any value which is to be altered is detected and the new value entered.

At all times during this program a message or messages are displayed to inform the designer of the currently available options. This, together with properly designed screen layouts enabled the program to be used before any instructional document for users was ready and has meant that the program is normally used without any such document being on hand. (Computer documentation for all the programs, i.e. coding sheets and flow charts, are, of course, fully maintained by the computer staff.) The program has been used many times for periods of the order of half an hour to an hour to obtain results that would otherwise have taken a week or more.

The second program is a very simple one concerned with the design of a "firtree root", which is used to attach a turbine blade to a turbine disk. This program consists of only two displays. Names and values as they are entered form one display. On completion of the input data the program checks the specified root for compatibility with design and manufacturing standards. Any violation of these standards is detected and an appropriate error message displayed together with a *suggested* way out of the difficulty based on past design experience. This is most significant because it enables the knowledge of experienced designers to be immediately available to those with less experience.

When an acceptable set of input values has been entered the designer can request the second display. This consists of a graph of calculated and permitted stresses together with weight and other significant parameters. By varying parameters the designer can find the combination which gives the optimum root. The designer can have printed a detailed stress analysis of any root.

This program is currently being extended to access information which will enable a designer to check simply whether the necessary tools are available for the manufacture of a new root. Choosing such a root will save money compared with one for which new tools are required and also mean that no delay will be caused whilst new tools are designed and made.

These two programs are clearly straightforward applications which could be successfully implemented on the simpler kind of interactive display referred to earlier.

Phase II

Based on the experience with the discrete applications already implemented it was evident that significant benefits could be obtained by the proper application of interactive computer graphics at Rolls Royce. The most obvious benefit being in the ability to reduce the total time necessary to complete a task. Further the later model of the 2250 permits a sketching facility to be used.

In aero engineering certain parts have always involved specialised procedures in their design and manufacture. It was natural that the design of such a part—the turbine blade—was chosen as the next task. The work of specifying and implementing the required console procedures and calculation modules is well in hand. It should be appreciated that the equivalent of many conventional programs are required, few of which existed at the start of the work. These programs are concerned with aerodynamics, stress and general mechanical design, and part of the work is ensuring that all necessary calculations are covered. When the work is complete a designer will require hours of 2250 time to completely design a blade. These hours will be spread over a period of days allowing discussions with technical specialists on the difficulties which invariably arise. Normally several blades are being designed simultaneously for each of which more than one possibility may be under consideration. It follows from this that the filing problem is far from being trivial.

In addition, in Phase II the basic requirements for the more general facilities needed in interactive computer graphics are being established. These include geometric modules and filing modules as well as standards to be followed in the planning and implementation of such programs.

Phase III

Referring to Figure 1 it may be seen that the basis of all the activities in mechanical engineering from concept through to manufacture is *geometry*. The aerodynamicist communicates his requirements and criteria, involving some geometry to the designer. The designer communicates the geometric information which he creates to the technical specialists for their analysis, he also communicates it in final form to the detail draftsman. The draftsman processes the geometric information and produces the engineering drawing which

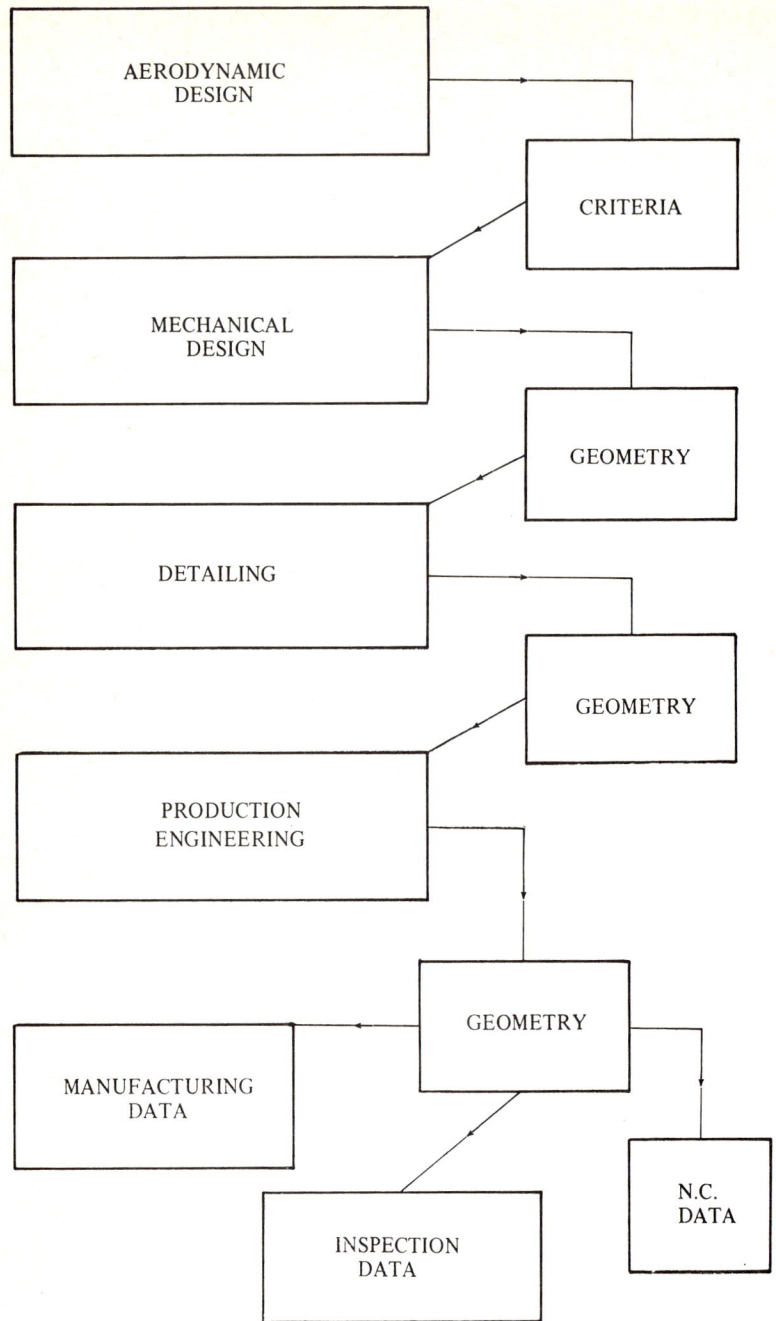

Fig. 1 Geometry's role in mechanical engineering

is sent to the production engineer as the precise specification of what is to be made. He in turn processes the information to produce further drawings as required for manufacturing and inspection.

I believe that by collecting the geometry as it is created and allowing each engineer in the chain to modify and add to it, using appropriate computing devices, significant savings in lead time can be achieved. The design of a turbine blade is a first step in that direction. In the next phase it is planned to extend the work onward from design and "sideways" to other key parts. In addition work will start on the general purpose procedures required to assist in the design and manufacture of other parts.

Summary

The currently available interactive computer graphic devices provide a natural means of communication between a designer and a properly programmed computer. This allows a designer to return to the earlier situation where he was his own technical specialist; without in any way diminishing the significance of the role of the specialist. Further, by means of appropriate programs it will be possible in the future to use computers for the generation, updating and storing of the geometric information which is the basis of mechanical engineering.

Biographical Note

B. T. Torson received the B.A. degree from Cambridge University in 1957. Since then he has been working at Rolls Royce Ltd. helping engineers and designers define their problems and formulate solutions, particularly in the field of mechanical design. He now heads the Computer Aided Design and Manufacture project.

Electronic Design Using Graphics

D. C. McDOUALL

Standard Telephones and Cables Ltd

Introduction

In this chapter I am going to describe the general approach which we have adopted in the application of computer graphics in our industry, and I will illustrate this with some selected examples.

I shall start by giving a brief description of our equipment and then describe the supporting software which we have produced. Applications of the facility will be seen most clearly in the light of these.

Equipment

We have installed an IBM 2250 Graphic Display Unit Model 1, with all the options except the Graphic Design Feature. This is attached to a 128K byte System/360 (model 40) computer, which is operated in multi-programming mode. A partition of 56K bytes is used, although this is shortly to be increased in size to 80K.

In addition to the display unit we have three techniques for the output of graphic data from the computer. These methods, which are complementary, make use of a specially modified line printer, of a draughting machine, and of a microfilm recorder.

IBM have made, to our specification, a special set of line printer characters for producing graphic output. (Character sets are readily interchangeable on IBM line printers.) The character set includes all the normal characters used by PL/1, in addition to short segments of line, arc, dots and so on which are used to make up a drawing. This method of graphic output from the computer is well suited to the production of all kinds of schematic diagrams, particularly those with a high content of alphanumeric information. The printer operates at a speed of some 600 to 700 lines per minute, and the software which we have produced permits drawings to be made up to 78 ins square. The drawings are produced page by page on the line printer, and subsequently butted together for reproduction by conventional dyeline means. The system is extremely flexible, the only non-standard feature in an installation being a facility for ten lines-per-inch printing.

The draughting machine which we have installed is a Gerber, a model 32 high precision table with the optical exposure head, and a 600 series linear control. The table has a drawing area of 48 in. $\times$ 60 in., a best accuracy of ± 0.0009 in., and a speed at that accuracy of 60 in. per minute. It is possible to remove the optical head and replace it by a pen turret. The machine is operated off-line, and reads magnetic tape.

The microfilm recorder we use, a Stromberg–Carlson 4060, is

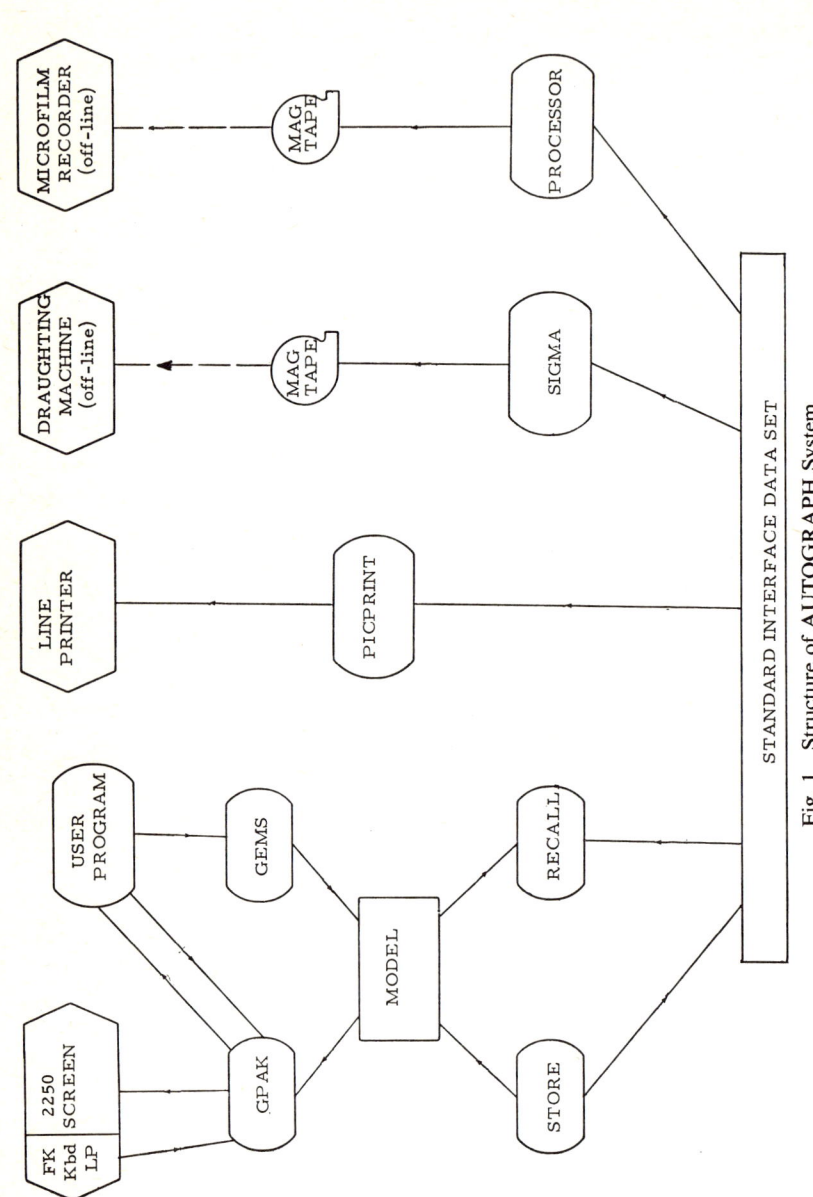

Fig. 1 Structure of AUTOGRAPH System

operated as a service by the Ministry of Technology. It reads magnetic tape off-line.

Software

The general structure of our AUTOGRAPH system is shown in Figure 1. Basically the concept is that of holding picture data in a standard format, and processing it selectively to a form suitable for a chosen device. At present AUTOGRAPH services the display unit, the line printer and the draughting machine. Extensions to include other devices will be implemented as required.

The Standard Interface file is a partitioned data set under IBM Operating System, and each member represents a picture. Users enter their own identifier for a picture which is then stored under an algorithm-produced name, and may subsequently be retrieved by the reverse of the process. Graphic data is held as types of element (horizontal and vertical character strings, symbols, lines, etc.) and co-ordinate data is held by a rectangular cartesian system with a range from zero to 16M.

The representation of the picture in core is by means of a model, and the model which we use consists of a direct access data set, on disk, with fixed length blocks. The handling of the model is carried out by our GEMS (Graphical Elementary Modelling System) package. The use of this simple method of holding the model, with sequential linking capabilities but with no ring structure, is extremely economic in its use of core storage. The GEMS package allows the manipulation of the model, the linking of elements in the model to members of other files or data sets, and scaling and windowing of the 2250 picture. In addition to GEMS, we use some of the IBM GPAK routines for handling the display.

The processing program for the line printer, PICPRINT, allows selected pictures to be retrieved from disc file, and reproduced on the printer to a scale which is set at execution time. The line printer is naturally unable to deal with all types of graphic data (for example arcs or lines at other than certain angles), and the program prints diagnostic messages if it is asked to process such elements.

The draughting machine processor program, SIGMA, operates in a somewhat similar manner. However, it has three additional facilities. One is the provision of optional "stop" commands on the tape when operator action at the machine will be required, and the second is the provision of a "move" command at execution time. This allows a single magnetic tape to be produced with the control commands for a large amount of output data. In this way it is possible to cover the whole drawing area of the machine with material and allow it automatically to produce a large number of items of output. The third is the facility to output selected parts of a picture, the parts being drawn superposed or separately.

It will be evident that while in principle it is possible to reproduce any picture on either machine, in practice one would not often do this as the two systems are designed for different types of work.

Applications

The uses to which this facility are put can be said to fall into two broad categories. In all cases information is presented primarily in graphical form, but whereas some tasks are essentially ones of graphic data processing, in others the computer is being used in conversational mode.

I will give two illustrations of programs which use the display essentially as a medium for conversation with the computer. We use the graphical version of the IBM circuit analysis program, ECAP, which allows operators to draw on the display unit a circuit of up to 100 linear, passive components and 25 nodes. The method of analysis (AC, DC or transient), is selected and the computer then analyses the circuit response. The results of the analysis are printed out, and the display operator has the option of calling for display on the screen the response at any selected node. Thus he may command the computer to display the phase of a particular voltage against part or all of the frequency range for which analysis is being carried out. In the light of this response he may then modify the circuit and repeat the analysis. Thus an extremely flexible tool for theoretical analysis is provided to replace much of the time-consuming bench testing work which has had to be carried out hitherto. While the ECAP program is somewhat dated now, it nevertheless still finds application.

A program which serves as a good illustration of conversational use of a display unit, is one we have produced for designing equaliser filter networks. In fact, the process is essentially one of using human skill in curve fitting.

The procedure operates as follows. The frequency response of a telephone landline is measured by the field installation engineer, and the figures are passed by telephone to the computer centre where they are noted on a form.

The data is then entered to the computer by means of the alpha-numeric keyboard at the display unit. It has been found that the advantages of flexibility and on-line data validity checking are considerable, and when the computer is operated in multi-programming mode the costs of entering data via the display are roughly comparable with the costs of entering it by means of punched cards. The response curve of the line is then displayed on the screen, as are also the response curves of the five types of filter which may be designed. This is illustrated in Figure 2.

The display unit operator then uses the light pen to select a type of filter, set parameters for the design in relation to the shape of a measured curve which is to be equalised, and command the computer to generate the filter design and analyse its response. This response is also displayed on the screen. The operator decides whether or not to

Electronic Design Using Graphics 173

Fig. 2 Equaliser filter design

accept this filter design, basing his decision both on the accuracy of fit as he can see it and on his ideas for other filter networks he will add subsequently. In this process the design of the filters is constrained by the requirements of the equipment practice. Thus, filters may not be designed which would require components of values that are difficult to provide or make. The program also checks for other essential parameters, such as maximum loss. Whenever the operator decides to accept a particular filter, the results of its insertion are applied to the overall frequency response, and the operator is able to see the cumulative effect as he proceeds with the work. Conversely, at any stage he has the facility of removing a filter which he has previously inserted and restoring the *status quo*.

The time taken to execute the program is variable. For lines without very severe distortion it is possible to design a set of filters in about 10 minutes, and frequently the final response of the line is better than is required by specification. However, some lines have a response to which it is much more difficult to fit equalising filters than others. This may be because of a very large spread, or because of a large number of sharp changes in response or because exact equalisation would demand components which it is not possible to manufacture. In these cases, the operator may try twenty or more combinations of filters in half an hour, until he is satisfied that he has achieved the optimum which the equipment practice will allow. When he is satisfied, the component values of the circuits are passed back to the field engineer.

Although this program is essentially geared to electronics work, the process could be applied in many other engineering environments. The main strengths of this method of approach to computer aid, compared with any non-interactive process, are four. First, it is cheaper in terms of development and, often, execution cost. Second, it allows for human foresight and experience in the entry of the data to the computer. Third, there is rapid and easy human control at every stage in the process, and fourth, it allows a man to disregard points in the measured response which he decides are of low significance, or possibly spurious. Alternatively, in the event that it would not be possible within the limitations of the equipment practice to design a set of filters which will produce a frequency response which meets a specification, the human can decide the manner in which the specification is not to be met.

The second type of application, as mentioned previously, is that in which the display unit and the attendant facilities are used primarily to process graphic data.

One example of this is in handling wiring data. Information about equipment which is placed in racks and is wired up is contained very largely in the schematic diagram of the equipment and in the associated stock list. Thus in entering the schematic to the computer, most of the essential data is entered. In addition to the schematic diagram, part of a sample of which is illustrated in Figure 3, other data is required from the computer.

Manufacturing data is required in various forms, for example wiring instructions whether for automatic machines or hand wiring, stock lists, and test data. Essentially, this is an illustration of a system where the computer process starts at the point where information volume is minimum. However, that information is best held in graphical form, and the process of transcribing into coded form for entry into the computer would be laborious and expensive. Thereafter the computer process is of the classical type, manipulating the data into the forms of output which are required for use. In point of fact one of these forms of output is graphical in nature.

Another example of graphic data processing is one in which precision automatic equipment is being used in order to achieve a high standard of accuracy. While the use of this equipment may not be the only justification for the system, it forms an essential part of it: the example I will give is the production of artwork for printed circuit boards. Once again, computer aid is applied to some extent in the classic manner. It ensures that design rules are obeyed, it processes the data into the different forms required for different uses and outputs it in that manner, turnround times are reduced, and so on.

Figure 4 illustrates part of a printed circuit board on the display unit screen. As the board is too large to be shown at convenient working

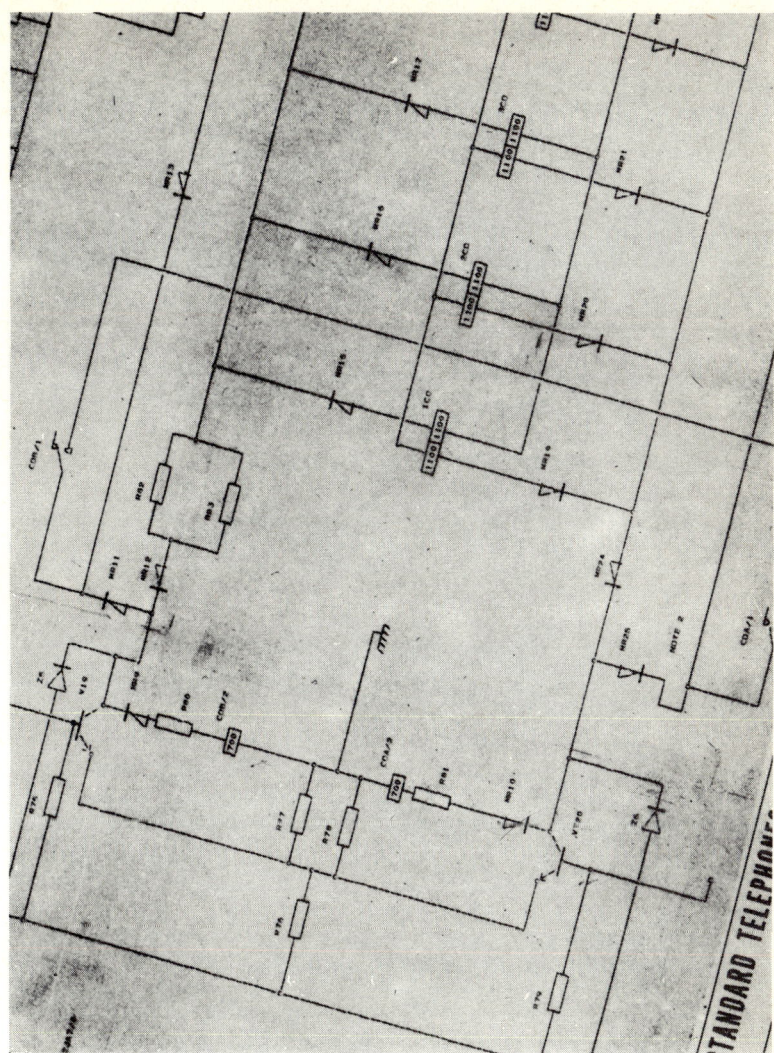

Fig. 3 Drawing produced on a line printer

scale, the screen is showing a window on the drawing. It will be noticed that the representation of the board has been simplified as much as possible: dots represent the pads, line widths are not separately identified on the screen (except by request from the operator). However, the accuracy of positioning is absolute, as the program allows only the resolution which is permitted by the equipment practice. Similarly, design rules such as clearances, allowable bend radii for component connectors, allowable designations of component and component geometry, are incorporated in the program.

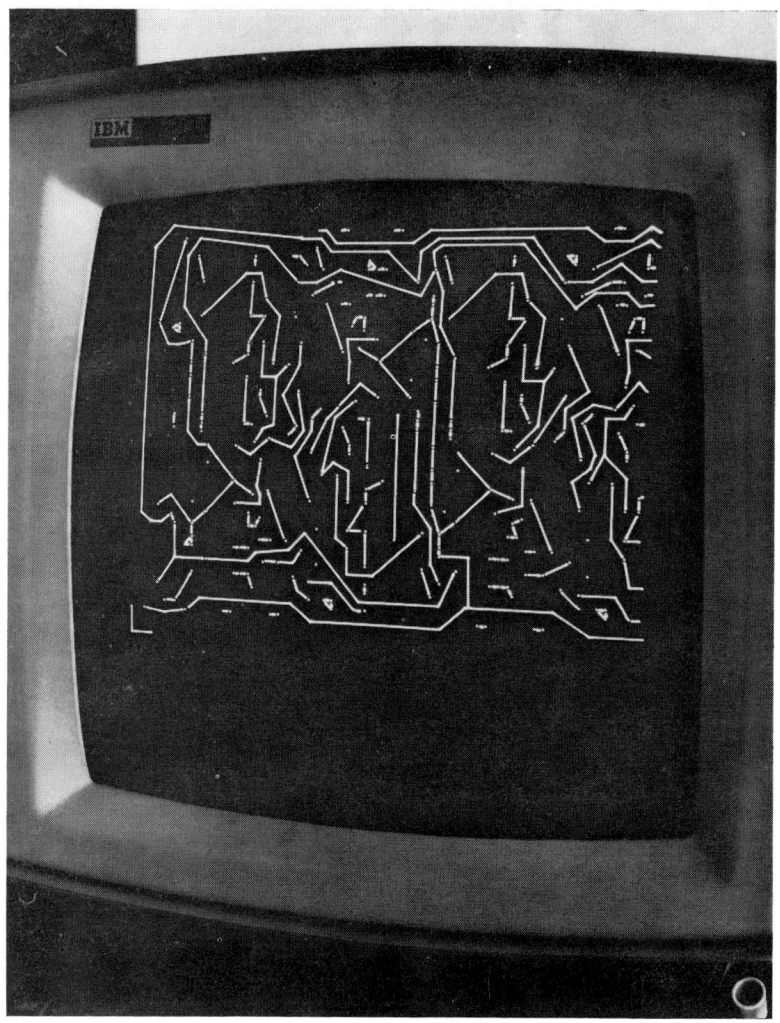

Fig. 4 Printed circuit board on display unit

Sufficient information for complete layout of the board is entered to the computer via the display unit. In those cases where updates are required to the standards allowed, these may be carried out immediately prior to work on the relevant board. When the data entry is complete the computer produces, either on the line printer or on magnetic tape, the necessary output for all documents and artwork for the production of the board. The magnetic tapes are then run off-line on the draughting machine.

As all designs are stored, either on disc or on tape, it is possible to recall earlier designs at short notice. When designs are evolving rapidly the graphic data processing facility allows fast and flexible response to meet changes of requirement.

I have tried to illustrate briefly the uses to which we put graphic data processing in Standard Telephones and Cables. To sum up: much information is more readily appreciated by humans when it is held in graphical form than when it is held in tabulated form. With the ability of a display unit to accept graphical data direct, and to redisplay it in such a manner that both man and the computer can comprehend it readily, it represents a very powerful medium for holding a dialogue with the computer for those jobs where this is the most appropriate method of working. Alternatively, it is a convenient way of entering graphic data to the computer if it is desired to process this to a variety of forms of output. This second application finds its widest use in those tasks where the information volume is at its minimum in graphical form, and the information needs to be processed to a larger volume in a variety of forms, including graphical ones.

Biographical Note
After reading Mechanical Sciences at Cambridge University Mr. McDouall underwent a Graduate Apprenticeship with the British Aircraft Corporation, at Weybridge in Surrey. Subsequently he became Technical Assistant to the General Manager and Vice-Chairman, and then undertook application studies of computer aided design.

In 1966 he joined S.T.C.'s Computer Aided Design Unit where he initiated the Graphic Data Processing Project.

Man-Machine Co-operation on a Learning Task

ROGER A. CHAMBERS

Research Assistant, Experimental Programming Unit, University of Edinburgh.

DONALD MICHIE

Professor of Machine Intelligence, and Director of the Experimental Programming Unit, University of Edinburgh.

Man-machine co-operation on a learning task

Artificial intelligence is the science of mechanising *interesting* brainwork. Most people would say that the really interesting kind of brainwork is not so much generating a solution to a particular problem but rather building a strategy for obtaining solutions in general. The use of computers in business is still largely at the level of helping the clerk, or at a more advanced stage, the designer. To be of significant help to the manager and long-range planner the computing systems of tomorrow will have to go beyond the manufacture of results to the synthesis of strategies. We further believe that the processes of strategy-building will be carried out by interactive man-machine systems, in which each partner has something to contribute which the other finds hard to provide. It is the long-term aim of our Department to produce an "intelligent machine" capable of interacting with man in this way.

The work reported here concerns our first primitive attempts to combine man and machine to solve problems. A feature of the work is the way in which the increasing use of graphic techniques has helped the development of a suitable interface between the two partners.

In the first experiment we investigated the use of a man-machine partnership on a task of network design—the famous Travelling Salesman problem. It is required to find a shortest route linking n cities on a map. Notice that here we are not concerned to generate a general strategy: each problem as presented merely calls for an individual solution in the form of a route. But this example of interactive problem-solving is useful as a prologue to later discussion of mechanised learning and strategy-building.

In this work (Michie, Fleming and Oldfield 1968) three ways of solving Travelling Salesman problems were compared:

(1) a human equipped with paper, pencil and straight-edge,
(2) a problem-solving computer program running in off-line mode and
(3) a human equipped with interactive computer graphics.

Each method was pitted against the same set of 5 randomly-constructed 50-city problems. Our panel of human problem-solvers consisted of 4 students. The off-line computer program used was the Graph Traverser (Doran and Michie 1966; Doran 1968).

The computer graphics facility consisted of a PDP-7 computer with a cathode ray display and light pen, together with a program enabling the user to connect points with straight lines, delete lines, print the total length of a route or part-route, and replace the current display by the previous best trial solution.

The results of these experiments showed that on average a human solver using computer graphics produced a solution six per cent better than that produced by an unaided subject, but the purely computer generated solutions were two per cent better still. On the other hand the additional cost of obtaining these solutions was great. The gains obtained by replacing unaided by interactive methods were, however, purchased more cheaply than the further improvement achieved by use of an off-line heuristic program. A noticeable feature of the results was the similarity of the solutions produced by different methods. Indeed the best interactive solution was often identical to the Graph Traverser's solution of the same problem and in one case the Graph Traverser was equalled by an unaided human solver. We had hoped that 50-point problems would tax the resources of both the Graph Traverser and the unaided human subject but this proved not to be the case. One of us (R.A.C.—not one of the panel) was able almost to match the Graph Traverser when working unaided. As we feel that man-machine interaction will be of most use in problems which impose great difficulty on either man or machine working individually, this experiment, though useful, was not able to fully explore the potential of interactive man-machine combination.

For some time prior to this experiment we had been studying a more difficult class of problem in which learning by trial and error is an essential ingredient. The problem, as posed, is learning to control a "black box". A black box is an object which has inputs and outputs, the connection between them being governed by some (unknown) law. It is required to experiment on it with sequences of inputs with the aim of arriving at a happy state of affairs in which the input sequence constrains the output sequence to remain within some specified bounds. Naturally when playing with this system we choose some unstable system from real life to put inside the black box. But the rules of the game require us to withold from the learning program, or from the man-machine learning combination, all knowledge as to the nature of the system to be controlled.

In the work reported here, the unstable system chosen to reside in the black box was the Donaldson (1960) pole-and-cart system, in which a motor driven cart must be kept running backwards and forwards on a straight track of fixed length in such a way as to balance a pole. The motor of the cart produces a fixed driving force and the only control is a switch which determines the sign of this force and so indicates whether the cart will be driven towards the left-hand or right-hand end of the track. The pole itself is pivoted on the cart, and can fall only towards either end of the track.

The apparatus is simulated by program rather than built in "hardware". In the initial stages of the project a learning program called BOXES and the simulated task were incorporated as the "controller" and "simulator" respectively of a single program running on the Department's Elliott 4120, or the Atlas. The design of BOXES, and earlier experimental work, are described elsewhere (Michie and Chambers 1968). In this program the simulator and controller pass information to each other in turn in a decision loop which is repeated every 1/15th of a second. Each time the loop is traversed the controller receives the current state-values of the simulated apparatus, calculates and passes on a control signal to the simulator which acts on that decision to calculate a new set of state-values to pass back to the controller. Since the significance of the numerical state-signals which it receives is not revealed to the controller, nor of the control signals which it sends back, the situation corresponds to a true black box in which the information available for piecing together a control strategy is kept to a minimum. Failure to keep the system within certain limits causes a "fail" signal which is noted by the program, and another run can be started.

With this system we found that in order to follow the progress of the BOXES controller it was necessary to develop a variety of monitoring routines which produced line-printer output indicating the internal workings of the program. This form of monitoring was useful but we were soon able to get a form of monitoring which was much more readily followed. The 4120 computer was connected by a high-speed link to the PDP-7 computer with display facilities mentioned in the first experiment. A simple display program produced a picture of the pole-and-cart apparatus on the display screen and by sending the output of the simulator to the PDP-7, as well as to the controller, the picture could be animated. This graphic display of the pole-and-cart apparatus was a great help in understanding the action of the controller (Figure 1).

The next stage was to make fuller use of the computing power of the PDP-7. All the simulator routines were moved to the PDP-7 so that the BOXES controller on the 4120 could work in parallel with the simulator on the PDP-7 with control and state-value signals being exchanged via the link. Since the simulator was governed by a real-time clock in the PDP-7, we were now able to simulate a real-time control situation which was fully open to human observation by means of the graphic display.

The next phase of the project was to introduce facilities to enable a human subject to undertake the role of controller rather than that of a passive observer. A control switch was made available by introducing

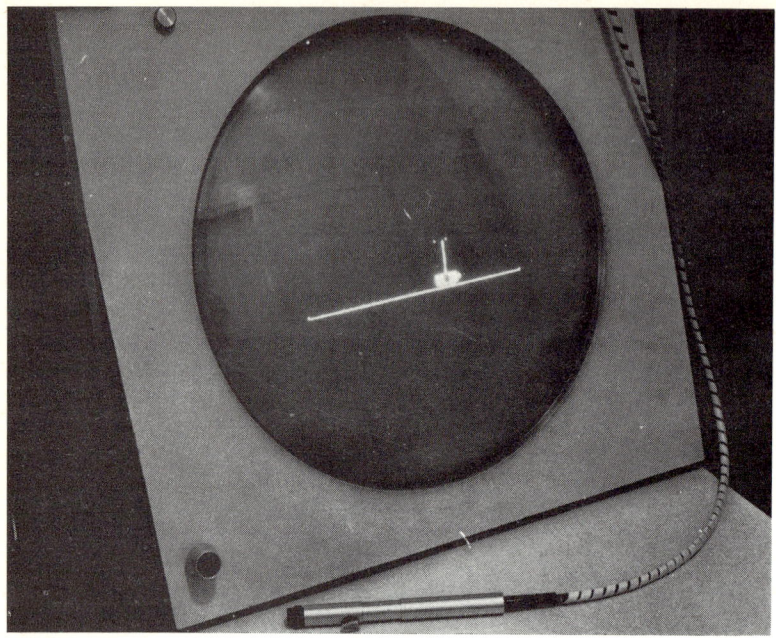

Fig. 1 Remote control with animated pole-and-cart display

into the display two "light buttons" activated by the light-pen. The manual control decision produced by this method could be incorporated into the system in several ways. In the present version of the "two-machine" system there are three modes of operation:

1. "Remote" mode: the light-pen is disabled and in every decision loop the control signal used is the "remote" signal generated by the BOXES controller.
2. "Manual" mode: only manual signals generated with the light-pen are accepted and if, in any decision loop, no such signal is available then the previous signal is repeated.
3. "Interactive" mode: manual signals are accepted when generated but in the absence of a manual signal the remote signal is obeyed.

In all modes the BOXES controller is informed at the end of every decision loop exactly which decision was implemented in that loop. Thus the BOXES controller is allowed to "observe" all decisions taken by any human controller, as well as its own.

As mentioned earlier the *significance* of the state-values and the control signals is withheld from the BOXES controller. This is still the case in the "two-machine" system which is further generalised to allow the "black box" simulator to be altered to simulate any system of up to four state variables by the re-writing of only two sub-routines of the PDP-7 program. To allow the display facility to be used with any "black box" a version of the program is available with a task-independent display representation of the problem in which each state-value is represented by the position of a pointer on a linear scale.

Figure 2 shows this display where the position of each pointer along each line indicates the value of the parameters such as the position of the cart, and its velocity, and the angle of the pole to the vertical and its rate of swing.

Using the scales-and-pointers representation of the control task, the human subject does in fact learn from experience. We at first thought that he would learn much faster than the BOXES program, and that the best use of the man as a partner for the program would be to provide a sort of "power-assisted take-off", improving his skill up to the highest level he could reach and then handing over to the machine to improve performance further. This idea turned out to be rather tricky to handle successfully. For one thing, humans do *not* learn faster than BOXES, at least not when the hidden task is the pole and cart. The difficulties occur in the first, interactive, phase of the experiment when BOXES is trying to form a summary of the human's strategy. In the second phase BOXES is in sole control, taking decisions based on past experience and occasionally making "experimental" decisions in an attempt to gather information leading to the improvement of its control strategy. However, in the first phase the initiative must lie with the human controller and the difficulties arise in finding

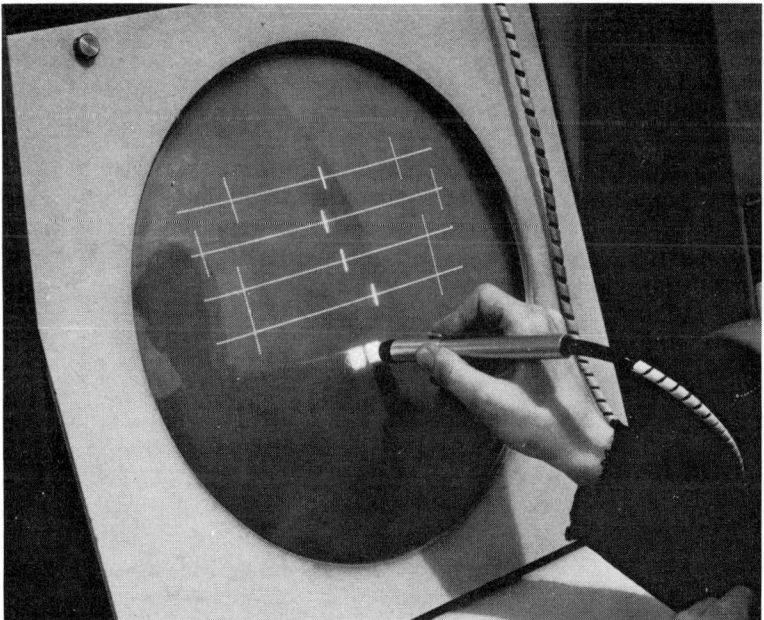

Fig. 2 Manual control with task-independent representation of the control problem. Each state-value is indicated by the position of a pointer on a linear scale

the correctly balanced partition of the roles to be played by the man and machine partners of the interaction. We can summarise the possible roles of man and machine as in Table 1.

TABLE 1 Four modes of operation in the Elliot 4100—PDP-7 linked computer study of "black box" learning

Mode of operation	When is the machine called on to give a decision
1. Machine learning by unaided trial and error.	Always, that is at each decision loop.
2. Human learning and transferring his skill to the machine.	Never. The BOXES program takes its decision from the light pen. If none is available, then repeats previous decision.
3. Fully interactive learning.	Whenever no light-pen signal available.
4. Advanced interactive learning	Whenever no light-pen signal is available *or* when machine's own preference has high confidence level.

First some remarks about the second mode shown in the Table. On the face of it, it looks rather dull, since not much is involved beyond a passive transfer to the machine of experience acquired by the human. We might therefore think that a machine that had been educated in this way could not outperform its trainer when left to its own devices. This is a mistake. The human's performance is variable even within a given situation, i.e. for a particular box, and one false step in a real-time task can be fatal. By contrast, the BOXES program which has accumulated for a given box the sum total of the human's experience will *consistently* apply, whenever it re-visits this box, whichever decision is indicated by the balance of this experience. Thus, if the machine makes the right kind of summary for itself of even quite low-performance human efforts, this summary can be made the basis of a much higher-performance strategy—the human's errors being "averaged out".

But the really interesting category of interactive learning is in the second half of Table 1. Here the challenge to the human partner is how to make the best use of a mechanical aid which itself has learning powers. One of our human subjects is working at this task, in session after session, and the performance of the partnership continues steadily upwards. In the process he is getting to know his partner's habits and susceptibilities more and more intimately, and he increasingly tends to delegate tactical work to the machine, himself intervening according to more global criteria. Whether by this means he will be able to overshoot the unaided machine and produce a control strategy quicker and cheaper is something we cannot yet say.

Nevertheless, graphics did give us a very quick and easy to assimilate method of monitoring our rather complex program. It also gave us a very effective interface between the man and the machine. Without some kind of graphics terminal, I don't think we could have done the experiment at all. Finally, it gave us an insight into the nature of the task, through trying it ourselves, watching the computer doing it, and watching a subject operate interactively. As a result of the insight, we were able to improve the performance of the program, and suggest a number of lines of attack for the future.

Learning to balance a pole is a far cry from learning to control a factory or a business. Yet the system we have described contains an embryo of decision-taking systems of the future. We must get used to the idea that computer aids to decision are going to graduate rapidly from mere digesters and displayers of data and become active participants in decision-taking, able to learn from whatever experiences their partnership exposes them to. By the same token the human partners will have to learn how best to steer the learning activities of their robot assistants.

We gratefully acknowledge generous assistance from the Science Research Council.

REFERENCES

Donaldson, P. E. K. (1960). Error decorrelation: a technique for matching a class of functions. *Proc. III International Conf. on Medical Electronics*, pp. 173–178.

Doran, J. E., and Michie, D. (1966). Experiments with the Graph Traverser program. *Proc. R. Soc. (A)*, 294, pp. 235–259.

Doran, J. E. (1968). New developments of the Graph Traverser. *Machine Intelligence 2*, (eds. Dale, E. and Michie, D.). Edinburgh: Oliver and Boyd, pp. 119–135.

Michie, D. and Chambers, R. A. (1968). BOXES: an experiment in adaptive control. *Machine Intelligence 2*, (eds. Dale, E. and Michie, D.). Edinburgh: Oliver and Boyd, pp. 137–152.

Michie, D., Fleming, J. G. and Oldfield, J. V. (1968). A comparison of heuristic, interactive and unaided methods of solving a shortest-route problem. *Machine Intelligence 3*. (ed. Michie, D.). Edinburgh University Press, pp. 245–255.

Biographical Notes

R. A. Chambers received B.Sc.(Hons. Mathematics) from the University of Newcastle-upon-Tyne in 1964, and the M.Sc. (Numerical Analysis and Automatic Computing) in 1966. Since 1965 he has been working as Research Assistant in the Experimental Programming Unit, University of Edinburgh.

D. Michie graduated in Human Anatomy and Physiology: M.A.(Oxon) in 1949 and D.Phil.(Oxon) in 1952. He has worked in the Department of Zoology, University College, London, the Royal Veterinary College, London, and as Senior Lecturer in the Department of Surgical Science, University of Edinburgh. In 1962 he was Special Consultant to the U.S. Air Force project for automatic language analysis at Indiana University, and Visiting Associate Professor at Stanford University, California. In 1967 he was elected to the Personal Chair of Machine Intelligence, and is at present Director of the Experimental Programming Unit at the University of Edinburgh.

Some Hardware, Software and Applications Problems

ROGER W. PROWSE
Brunel University

Introduction

At the Brunel Symposium on Computer Graphics in July 1968 the discussion time gave an opportunity for many experienced graphics specialists to ask and answer questions and to comment on some of the problems facing graphics users.

I have made a selection of some of the topics which were discussed where the information that was elicited might be relevant to a wider audience throughout the world.

Questions answered by Samuel M. Matsa, IBM, N.Y. (See Page 1)

A. W. Nicholson (Cripps Computing Centre, Nottingham University)
Q. You mentioned that it was possible to save time in doing a job using graphics, but you did not mention how long it took to program the operating system to make the job possible.
A. This is a big problem and is more than just a graphics problem. In order to have a good practical graphics application, one needs a very good operating system, at least a multi-programming environment, perhaps a complete time-sharing system, so a great deal of time goes into software preparation. However, in the simpler cases it need not be considerable, since one uses a dedicated system and one can convert operation to the graphic rather than the standard batch processing mode comparatively simply.

D. A. Franklin (Medical Research Council)
Q. What is the minimum size of computer that you would expect to be feasible for a graphics system? Could you get anywhere with, for instance, a 32K, 24 bit word length computer?
A. Yes, but it depends upon the application; it is impossible to answer the question in general. There are applications that could be implemented on a 32K machine, assuming that you have some additional storage, preferably available on a disc, or on a drum, to hold all your displays and images. There are, however, other applications that would require much more computing power than you would have available on the medium size machine that you mentioned, and there are also people doing useful graphics on even smaller systems.

M. I. Bernstein (System Development Corporation)
Q. Would you enlarge on the problems of incorporating interactive real-time displays under a time-sharing system?

A. I shall comment in terms of the experience I've had in using graphics with the M.I.T. time-sharing system. One of the important features is as follows: If you are working in a time-sharing system with graphics, it is not always possible to respond immediately to the user, but it is essential to have the operating system, i.e. the supervisor that is always in charge, to indicate to the user that when he pointed with a light pen at something, or when he pushed a function key, then the system recognised this action.

Questions answered by D. R. Evans, Royal Radar Establishment (See Page 7)

T. *Daniels* (Daresbury Nuclear Physics Laboratory)
Q. You mention gas discharge tubes as being a possibility. What sort of resolution do you think would be obtainable from these?
A. Not very good. At the moment we get about 16 elements to the inch, so that you will get a 16 by 16 per square inch device. It may be possible to improve on that, but at the moment that they are extremely experimental. There is really a double struggle going on in the field of computer graphics. There is a tendency to want large amounts of information on display, but we think that one of the things we have to do is to reduce the amount of information to be displayed. By using the computer to pick out the problem we are worried about and not the total problem, it is possible that we will not always need the very high resolution device.

A. W. *Nicholson* (Cripps Computing Centre, Nottingham University)
Q. In dealing with the conventional solutions I think you have missed out what I believe to be the best solution of all, which, instead of having a local memory, is the small computer. These are mass produced and are therefore much cheaper than making a special memory with all its test equipment, particularly as the small computer can act as the test equipment.
A. You invariably find in an applications field that you always start off with a small buffer store, but very rapidly it becomes a small computer to carry out editing and local processing operations. The small computer is usually the solution. However, the small computer invariably starts to do more and more jobs and may become not such a small computer. It is inevitably expensive, and this is a big problem. However, a small computer can make a very efficient graphic display, useable at the end of a telephone line, but what is needed is such a computer at even lower cost than those at present available.

R. *Guedj* (Compagnie Internationale Pour L'Informatique)
Q. Could you say something about circle and arc generation on displays.
A. In general, what everybody tries to do is to approximate to the arc as best they can, by drawing a series of short straight lines. This takes

time. You can also use function generators based on Lissajous type figures to develop arcs as you want. This means that you have to include in the system a series of frequency generators. These are then combined on the tube face with sufficient bright-up signals, by chopping this one up and adding that one in (doing electronically what you would normally do with a pencil and a series of French curves), to give the required curve. This works, but is a more expensive solution.

Professor C. W. Gear (University of Illinois)
Q. If a customer had a specific single application for computer graphics, for example, numerical tool control programming, do you consider it would be better to couple a set of display systems directly to the central computer or to couple the several display systems to separate local memories and couple those to the computer?
A. It depends entirely on the following. In general directly coupled displays would be a cheaper solution but would place restrictions somewhere in the system. The local memory is undoubtedly the most flexible solution and would leave the systems engineer in an easier position, but it does depend on the application. We ourselves have, not a tool-room problem, but an air traffic control problem, which we started with directly coupled displays. Now, as is inevitable, the customer changes his mind and wants more and more information. The net result is that the refresh rate is dropping so far in this system that though it's cheaper it's now becoming unacceptable. So we were driven to using local memories. As always, if you have the money you can have a more flexible solution, if you haven't got the money, make sure you have what you want.

Questions answered by S. Bird, The Marconi Company (See Page 17)
I. George (I.C.L.)
Q. Can the programmer declare blocks of his own type and add data of his own type to the blocks and add associations to the blocks?
A. Yes. This we envisage. There's no reason why he shouldn't do it at the moment in fact, though only at low level. Across the interface between the graphics package and the applications package could be passed verbal descriptions describing for instance the nature of the ring structure. This would enable the high level program to search in or build and modify the ring structure. The problem arises in that although you may say you are working in a high-level language, in effect you have got to develop yet another language, consisting of a set of procedures based on that one. I don't know of any standard method in high level language for describing what you want to do to ring structure.

We developed this ring structure for display purposes, but it is general—there is no reason why it should not be used for other purposes such as generating the problem model. The display and the problem program could then work on the same model, which could be very convenient.

A. G. Price (Documentation Processing Centre)
Q. I understand that the Marconi Display System gives facilities for either absolute positioning on the screen or positioning relative to the last point, but that there is a risk of drift in the latter case. You specify in your list processing structure that you can have relative co-ordinates. How does this interact with the actual mechanism for display on the screen? Do you find that with relative co-ordinates you have to recalculate absolute positions from time to time to keep your display from drifting?
A. The answer is that the only part of the program that worries about this sort of thing is the very last lap of the display program. This is fed with various parameters about the display, such as how long it can go on without being repositioned and this program automatically puts in repositioning where necessary. In this sort of system an actual display file is virtually a trivial end product. If we are repositioning a sub-picture then this sub-picture as it appears in the display file may contain many repositioning words. These will have been put in at the necessary intervals by the display program and these will be updated automatically by the display program as the sub-picture is moved.

J. H. Ludley (National Engineering Laboratory)
Q. Have you found a reasonable method of splitting up your structure between main store and backing store?
A. We have not actually taken that step yet, although we have laid the ground so that the programmer writing ring handling routines can assume infinite core. This is done by a method which tags points where rings go across record boundaries, or across the boundary from core to disc, and we have done this in a way which in no way slows down the speed of operation if everything happens to be in core. If the programmer calls a routine which says "Go to the next item on this ring", then the program will go to the next item on the ring. If it has to do a lot of transferring to and from disc to do it, it will take longer, but that is the only way he will know.

Questions answered by A. R. Rundle, Elliott-Automation
(See Page 29)

S. Bird (Marconi)
Q. To what extent does the type of input program of figure 5 (page 34) have to be re-written for different applications?
A. We are still experimenting. Though a lot is duplicated between programs, there is enough that is not, for us not merely to produce these functions as standard packages, and put them into our software.
It may be all right for drawing lines, but with real problems, these

lines may represent pipe runs and have coefficients of expansion and thicknesses and a lot of other data to be incorporated as well. The same applies to buildings, when analysing extensive frameworks.

So in a real problem, the user builds up his data structure, and compared to that, his task of writing graphic input programs is fairly trivial. The important thing for the manufacturer is to provide the basic subroutines for graphics and for data structure, so that the user can build up a data structure that is really useful for his application. We are not therefore providing a range of complete application programs which at present would be of only limited capability, and hence restrictive.

Anon
Q. Are the difficulties in positioning the tracking cross when doing freehand drawings due to hardware or software constraints, rather than to human limitations?
A. Yes, basically hardware constraints, because one works on a grid which gives typically 0·01 in. between each point. The light pen is fairly blunt and one has no ruler or set-square, though given time, the pen can be positioned quite accurately. Several things can also be done to help the user. The exact pen position can be displayed and, if you take time, you can probably position it accurately to the one increment on the screen. I must emphasize that this generally takes time and the whole idea of using a graphic device is that you want to cut down your time. So, basically, it is limitations of the hardware, although you can get round them. Obviously, to draw accurately, the easiest thing is to expand the bit of picture being drawn, so that inaccuracies owing to hand wobble, etc., are irrelevant.

F. M. Larkin (UKAEA Culham)
Q. Should the data structure reflect something of the quantities in the model which one requires to be invariant under picture modifications, for example, metrical, projective or topological properties? If so, are there some unifying principles that apply to a range of data structures or is every programming problem completely distinct?
A. The way in which you build up a data structure depends entirely on the task. If it is important to know that a particular object occurs in many forms, in many places, on the screen, then there must be some means of storing this information. If, on the other hand, there is a straightforward array of lines connected between two points, they can be held in arrays without a proper structure. Where you want to draw in three dimensions and you want to build up the whole out of components (an example might be pipework, or three-dimensional frameworks) then you could probably arrange a standard formatted data structure, but in general at the moment the problems are so different in each case that it is not worth while providing a standard data structure. We are trying to make it as easy as possible for you to

generate your own structure. Time and experience may change this; we may decide that it is worth while restricting you to give you easier use of the display. It is difficult to say.

Questions answered by Harry H. Poole, Raytheon Company (See Page 41)

M. I. Bernstein (System Development Corporation)

Q. Would you enlarge on the area of character sets, that is the number of characters and the character set available and character size and character style?

A. Regarding the number of characters, there are some applications which can get by with 32 characters (which is 2^5) but most applications require 64 characters which allow you to have the 26 alphabetics, the 10 numerics, a series of possibly 8 to 10 punctuation marks and then about 20 special symbols. In the United States at least they've been going to the American Standard Code for Information Interchange (ASCII) which is a 7 bit character system which allows up to 128 characters. They find a lot of applications require this for extra control characters and special characters. There is also a 6-bit sub-set available. As far as the sizes of characters are concerned, depending on the size of the C.R.T. and the distance of the operator from it, there is a lower limit to the size that any operator can see clearly, and this may be 3/16 in., $\frac{1}{4}$ in. or 5/16 in. It will be typically in that range for most applications. Starting with that as a low range, then for the operator to distinguish the different sizes, if the smallest character is $\frac{1}{4}$ in., the second size character may be a half inch and then the third character size would be an inch. That is about the range of character size that you can typically put on a screen before the top size starts cluttering the screen so much that it becomes wasteful. Finally, about fonts. There are many types of character generator and for low cost systems you only want information for the operator, so the basic criterion is readability. There are other systems for direct printing which need very high quality character generators and these are also available. C.R.T.s of comparable quality must also be used.

E. G. Chadwick (Hawker Siddeley Dynamics Ltd.)

Q. Your remote display system seems to be purely for transmitting graphic symbols, numbers and letters. What are the possibilities now of displaying remotely graphs and line diagrams, etc.?

A. You can transmit remotely line drawings as well as vectors and characters. The only problem here is that the display at the end of the line has to be a lot more complicated and therefore a lot more expensive to be able to handle vectors in this way. Since it is remote information you also generally have to do the refresh locally instead of in the computer, which again adds to the display hardware cost and so forth. Except for a bandwidth problem, which may limit how often you can update this information on the display, there are no fundamental differences.

G.G.J.A. van der Eycken (*Philips Computer Industrie, Holland*)
Q. We have heard something about the human factors of the light pen and function keyboard. I wonder if you could also tell us something about the use of the rolling ball versus the joystick.
A. The subject is difficult, but the two are about equal and it is largely a matter of choice. Both have problems in maintenance, in accuracy, speed and so forth and for every advantage, whether it is human factors, or electrical design, or programming, there is a corresponding disadvantage.

The track ball is generally used for tracking devices where you are trying to keep a cursor on something which is moving, but, except for that small distinction there is little else. I have just finished a study for the U.S. Army, the net summary of which is that "you pays your money and you takes your choice". In general, the people who are trained first on the track ball, like the track ball, and vice versa. There are a few exceptions to this, of course, but in experiments to obtain some analytic data instead of subjective opinions you find that the two are essentially equal with the minor exception of the very fine tracking tasks I mentioned, which are very rarely used in this kind of application.

Questions answered by C. Machover, IDI (See Page 61)

Dr. R. E. Thomas (Atlas Computer Laboratory)
Q. There is the possibility of displaying lines on a screen for which the light pen is automatically disabled by unsetting a couple of bits of the display word. Could you say something about the usefulness or otherwise of this feature? The DEC 340 for instance has this facility.
A. Typically, in any of the more complex computer graphic systems, part of the control display format has control bits which cause the light pen interrupts to be sensed by the central processing unit and the programmer has the option of masking blocks of data. Thus even though there may be electrical output from the light pen, the program ignores these. For example you could put grid lines as background, and though the light pen may be in the area of the grid lines, it would sense only data over the grid lines that had not been masked by program control. In a complex system, this is a very simple facility to include.

R. E. Rosing (University of London)
Q. I am a geographer and I work with topographic maps and over such maps we are faced with a tremendous problem of data input. Very little has been said about the problem of graphical data input. We are currently faced with the problem of having on the order of seventy-five million individual bounded districts, in other words, an infinity of irregular polygons contiguous one with another. How can this sort of thing be scanned, digitised and somehow made machine readable?
A. This is the traditional question asked by people who are looking

at the question of computer aided design, where they have stacks and stacks of engineering drawings and would like to start a new drawing program. What do they do with all their old drawings? Can they digitise these, put them into their files and thus create a data base from which to operate? The answer is that one can, but the economics say that it is probably not worth while. One decides to start from the present plateau, and build up a new data base. The digitised data entry that you describe is also a traditional problem in seismic research and nuclear research and techniques have been developed there for high-speed digitising, but by any standards any chart you draw is relatively expensive.

There is a company in the United States called Information International, which is specialised in the area of very high-speed digitising. Their work is associated with nuclear research, and their systems typically cost in the order of three hundred thousand to half-a-million dollars. It is the same kind of magnitude of some of the hard copy systems that first came out with the IBM 360/2250. The question, of course, is speed. There are relatively slow mechanical ways of putting in data, and many of you may be familiar with plotting devices, in which you position a cursor over a point of interest, press a button and it will then punch a card which has the digital information on it. If you have ten points this is practical, if you have millions of points, it is very impractical.

The oil industry has essentially the same problem. They have a long history of charts they would like to digitise and operate with in a computer based system. Not only do they have the problem of a tremendous amount of data, but they have the problem of separating the chart data from the background which may be overlaid with grids. In some cases even, the charts are sufficiently old that they are simply dirty and the question is to digitise the information and not the fly specks!

Though in theory it can be done, in practice it is a terribly expensive process and, unless you can justify the investment, you might be better working in the interim with a dual system: a traditional storage system for the old data, with new data going straight on file.

Professor B. Herzog

Comment: It is worth noting that in the area of high energy physics and biology, where a lot of photographic data has been obtained from experiments, two special computers at the University of Illinois, known as the ILLIAC 4 and the ILLIAC 3, an optical scanning computer, are largely dedicated to hardware implementation of the digitising process, to clear the tremendous backlog of work built up of recent years. However, if you have got all that data, where are you going to put it? The large data base requirement is still a severe economic problem despite predictions that the computer is going to take on all the routine work in these fields.

However, if we discourage you with such comments, we will do ourselves and you a disservice. I think there must be a transition period where people will use the computer for new data, and co-ordinate that with traditional methods for the old data. In time there will be more data on file available for the regional planner and the geographer to use in a comprehensive way, but I do not think that we can expect to say when the change will be complete.

I am trying to encourage the people with the problems to try some of these things, because I fear that some of our guesses as to what the future holds must be wrong, through ignorance of what the geographer, for instance, wants to do. Do not rely only on the computer people to tell you what it is all about.

Question answered by F. E. Taylor, N.C.C. (See Page 85)

C. C. M. *Parish* (Central Electricity Research Laboratory)
Q. Is it true that most of the money spent on graphics is going into systems like BOADICEA, and those of the power industries?
A. It depends upon the area we are talking about. In terms of applications, it is true. A vast amount of capital is going into airline seat reservation systems. In terms of hardware a certain amount of money is going into developing high resolution tubes, better co-ordinated displays on so on. In software quite a lot is going into making possible the use of displays with high level language statements.

Questions answered by Murray A. Ruben, D.E.C. (See Page 91)

B. L. *Wood* (I.C.I.)
Q. One of the most annoying things that happens when an operator is using a display tube is that he gets a rather bad flicker on the picture especially when there is a large amount of data on the screen. As direct view storage devices seem to have a very high information density and also a completely flicker-free picture, do you think that in the future, if the response time is speeded up, the storage tube could take over completely from the refresh system?
A. What you are proposing is a considerable advance in materials science, which I do not see on the horizon at the moment, but certainly, if this were the case, one would have a good argument for proposing that a storage tube take over altogether. When you look at it from a systems point of view, it solves the problem that you have. It stores the information where you want it. You do not have to transform it from some other source, which is what you have to do in other systems.

R. I. *Macdonald* (U.K. Atomic Energy Authority)
Q. How long can one expect a typical storage tube to last? Do they last indefinitely?
A. The specifications that have been stated by Tektronix are approxi-

mately two thousand hours of life at full intensity. These numbers are stated only on a theoretical basis at present. There has been one system which Mr. Stotz of Computer Displays Inc. has designed and this has been in operation since last summer, and tends to confirm the 2000 hr. figure.

Questions answered by Prof. C. W. Gear, University of Illinois
(See Page 109)

K. J. *Durrands* (Vickers Ltd.)
Q. I find it very difficult to determine how you decide when you have got a number of engineering complexes which are spread some hundreds of miles apart how big your remote terminals should be, and how much work should be left to the main computer. I should like some views on this considering that we have certain service computers belonging to the Ministry of Technology and the National Computer Centre.
A. One needs to examine the class of applications. If we are talking about a system that inherently needs multi-person access (a hospital information system, for example), the terminals will probably be passive, with no computing capability and they can be low speed graphic or alphanumeric C.R.T. devices. If we are talking about more complex computer aided design problems, the answer must depend in part on whether all users of a given central computer will be working on the same small class of problems, or whether they will be working on totally dissimilar problems.

In the former case, software and hardware placed at the central computer can be shared, that at terminals cannot. In the latter case, the key question is often one of communication bandwidth and cost. At our present level of understanding, I would advise a potential user not to consider this problem at all, but to concentrate on developing the application system using a C.R.T. display local to the computer. When this system has been completed, it can be analysed to determine the economic feasibility of making part of it remote. The problems of using graphics are still sufficiently difficult that I consider it unwise to muddy the waters further with communications problems.

Professor B. Herzog (University of Michigan)
Comment: I should like to endorse that vehemently. The experience of the Ford Motor Company where I held a position for three years was that it was extremely useful to get a typewriter entry time-sharing system because they are, in fact, exactly in the position that you indicate—disparate locations not only in the Detroit area, but also nationally in the United States, and abroad. The effect of this time-sharing computer system was to bring computing to the engineering force. The problem of graphics is far more difficult and General Motors, who have been leaders in this particular area, are trying to face that problem by putting 2250s at large distances from their central com-

puter, but "large" is still less than miles. There are problems, and I think that there are stages in which to approach this thing that will still meet the need of the engineering complex. I think the answer is to use remote typewriters, with the graphics terminals local to the computer, so that one can get something started, while waiting for a full system of remote graphics terminals.

M. I. Bernstein (System Development Corporation)
Comment: Concern about the cost of character recognition programs has been expressed, and is partially justified, but for the last year-and-a-half we have had a character recogniser working under the existing time-sharing system at S.D.C. This is on the Q32, which makes it a unique system. It is not easy to transfer data about that system into terms of, say, a System 360 machine, but the concept is that the character recogniser will be a system program and, therefore, occupy the overhead portion of core and execution time and the user's only required extra space for character recognition in his program will be a dictionary of approximately 1500 words of 48 bits.

We also are able to get reasonably good response time even under the time-sharing system by, instead of allowing single character input and waiting for it to return, using multi-character input. Therefore, the response time allowed can be longer than instantaneous without getting the user particularly upset. If he writes in five characters and gets five characters back with a short delay, he feels better about putting in one character and waiting the same amount of time to get one character back. Part of the system problem, because of this remote interface, is to supply enough buffer space to let him write the amount of input that it requires to describe several characters.

Professor C. W. Gear
Comment: Do you have any figures for the amount of C.P.U. time used to recognise a single character?

M. I. Bernstein
Comment: The times I have are stroke times as we process a stroke as the basic unit of information. A stroke is from the time the pen goes down to when the pen comes up again, so a character may have up to five strokes. I do not think that there is anything anybody draws that takes more than five. Our average C.P.U. processing time on the Q32 to process a stroke is 16 milliseconds. This includes all the analysis in the dictionary lookup for that stroke and there is the statistics gathering that obviously goes on, included in that 16 milliseconds.

In order to give you some idea of the work possible in the 16 milliseconds, the Q32 has a basic core cycle time of about 2 micro seconds, has a partial look ahead for the next instruction and has a reasonably more powerful instruction set than the average computer in that we can perform dual arithmetic on coordinate information: that is we can perform simultaneously the transformation on X and Y, such as add to, subtract from, multiply both parts by a constant, etc.

D. S. Hutchinson (British Steel Corporation)
Q. At a time when in system cost the programming percentage is going up, it is refreshing to hear from Professor Gear that some attempt is being made to reduce this; however, I feel his present system is rather sophisticated for some real-time applications where something much simpler such as editing, debugging and testing assistance to programming would be of very much use. I believe that on some systems, notably C.D.C. and IBM, techniques for using C.R.T. displays to help in editing and testing programs have been implemented. Could Professor Gear tell us what reduction in overall programming time can be expected from such techniques?
A. I qualified my talk by saying that this was a research project. In other words, it would currently take longer and be more expensive with the methods we are using. I would not expect anything out of this type of approach for a number of years yet, and I frankly have not seen any approaches to the use of flow charts that have had too much impact, except possibly in the documentation area, where some of the now available, proprietary software programs that produce flow charts from FORTRAN and assembly language programs are having some effect. AUTOFLOW, for example, has been used at the USAEC Argonne National Laboratory near Chicago. When it was first made available, and "advertised" to their users, it was used six to eight times a day. After about eight weeks the advertising was stopped and use dropped to once or twice a week. However, this may indicate that it is being used mainly to provide documentation for finished programs.

Questions answered by A. E. P. Fitz, Navy Dept., Min. of Defence
(See Page 119)

J. R. A. Jones (Admiralty Underwater Weapons Establishment)
Q. I should like to ask about the interaction between design problems and business problems. We have seen quite a lot on computer aided circuit design and one can foresee the design of quite complicated circuits within a few minutes. We could possibly produce the diagrams showing the circuits also in a matter of minutes. The next requirement for the designer is to know whether the particular transistors or resistors he requires are actually in stock, and there is little point in his now spending another ten or fifteen minutes walking across to the storehouse.

Does Mr. Fitz think the designer can have access to his stores file? Secondly, is he prepared to give us a reservation period which is a little longer than thirty minutes, because one does not want to draw the things out, but just to make a reservation on these items at this particular stage? Finally, if the designer knows what is in stock, then he can bend his design to suit the stock. This may well be an inferior

design and, therefore, would the speaker accept information in addition to the demand information, from the designer to say this is what he really wants, but he is prepared to take what there is in stock. This will largely prevent a file system starting whereby stocks are up-dated and further demands are made for a stock simply because somebody ordered the stock in the first place.
A. May you have access to the stock file? Yes, why not? If you want to get access to the mail order depot which provides them, then there is not the slightest reason why you should not ring up and ask about the resistors you want. The fact that they have enquiry terminals makes it easy for them to answer your question. However, I do not think that the fact that we have a great big surplus of those resistors, for example, ought to influence your design at all. If, in fact, you will need them, there is no reason why you should not write in and ask for ten thousand to be earmarked for you, provided that you ask for them to be earmarked, e.g. for a month, and you remember at the end of the month to say if you do not want them, or need fewer, or something like that. The great difficulty we are in always is that technical departments come along with material estimates which are forgotten very soon after they are put in and one of the reasons for our pre-vetting of input is to try to filter some of these requests and relate them to former material estimates. Hence, I would be chary about giving enquiry terminals to all and sundry and allowing them to put long-term earmarks on our stock. Like any other business, we have to watch how much capital we tie up in dead stock.

With regard to your third point, if you are designing, use the most efficient component rather than the one we may have in stock, but remember to tell us that you are doing this in time for us to get some. This is a common problem for most ordinary organisations that a new product comes out and there is no support for it because the designer forgot to tell somebody that he was using a new component which is only made by one factory and cannot be bought off the shelf. Equally, if you decide to use shelf stock but would like something better ordered for the future, please tell us what you want in good time.

A. W. H. Carter (Plessey Radar Limited)
Q. I should like to continue with this question about whether designers should make use of components in stock. I agree with the reply, but there is something the designer can do here and I think the system that the speaker has outlined in his paper might lend itself to it. If I could hark back to the TSR2 aircraft to give an illustration, in TSR2 we had a problem of deciding how we could best implement reliability. We looked at all the transistors we used then, and there were well over one hundred different types. If we imposed reliability tests and quality control on all of those, it was going to be extremely expensive. However, we did find out, by getting component population counts (a complete analysis of all the components that were being used) that only

twelve types of transistors covered 80% of the transistors used. Thus, if you have got the information in your computer which says these are all the components in use, you can tend to get preferred components in a very practical sense. In one sense you can get a non-practical preferred component, which is a standards man saying these are the components you should have. On the other hand, you can get stock lists saying that these are the components most commonly used. If the designer can be provided with this information, then nearly always he can tailor without any penalty on his designs into these preferred types. Once you have done this, you are in a much more controlled situation. So I think there should be feedback of components being used, to the designers, to help rationalise the designs. I would be interested in your comments.

A. I am interested in your question, too. There is a preferred list of all components in the Admiralty, but it is difficult giving in period printed stock lists the sort of feedback that you require because stock is constantly turning over. The object of a stock control program is to minimise stocks, to keep the capital investment in stocks as low as possible, and as you suggest, I think the answer to this lies in coming in on our enquiry network and obtaining stock turnover figures. Then, if you can use a line which is always in stock, there is a trade-off for you and for us.

A. F. Nightingale (Vickers Limited, Barrow)

Q. This question to some extent leads on from where you finished. I am rather more interested in the estimating side of design. I think designers have a tendency to do the design and give the complete design to the estimators, who then say this will cost X pounds. I think the design could often be cheaper had the cost been taken into account and the design been optimised on the way through. Are you going to have a feedback from your stock control, which I imagine carries all this costing information, to the designer at a stage when he can formulate his design using this information?

A. You are putting to me a value engineering question more or less. I can only answer this by saying that, if you, as a designer, can have access to the system, you can find out not only what stocks are held, but also what their cost is, but the one snag to all these ideas is that one designs a very nice little in-house computer system which does the business of the area concerned, you lay on an enquiry service for all the people there and then people outside that area want to come in and the system tends to mushroom. As with many other computer projects, the computer gets overloaded and sinks under its own weight. It is a splendid thought to give everyone access to all the information, but certainly in our own context I do not think that it will be really practicable until we buy another machine, and this one has not taken on all its other jobs yet.

Questions answered by P. E. Walter, West Sussex County Council
(See Page 125)

A. E. P. Fitz (Ministry of Defence)
Q. Did the software for your architectural application come as a package or did you write it yourself, and if so, what effort was put in to it?
A. We wrote the software ourselves: myself, an architect, with two quantity surveyor colleagues, and I would acknowledge the invaluable assistance of our IBM systems engineer. We started in February of last yeat at which time we did not really know what a computer was. The program is written in FORTRAN and uses an IBM package called GPAK which allows us to draw lines, points and characters, etc. Our program makes use of this to draw shapes as required.

A. G. Price (Documentation Processing Centre)
Q. Mr. Walter mentioned that data from the graphic system was fed into an existing bill of quantities program. Did all the system programs get designed together or did you have to fit programs together which were not actually designed as a suite?
A. No. They were not designed as a suite. The bill of quantities program to which I referred has been used by my Authority for some two or three years. The quantity surveyors provide the data which is punched on cards and thus submitted to the bill of quantities program. What we have done is to replace the cards with a 2250. In fact, special areas of the building which cannot be described via the display are still treated traditionally. This data is merged with that from the 2250 and goes into the bill of quantities program. There is no difficulty with compatability between the two suites.

P. A. Purcell (Royal College of Art)
Q. You said that if the design is satisfactory it is then transferred to the magnetic tape, but if it is not satisfactory—if some modification or optimisation is required—what procedure do you adopt?
A. We have the facility of recalling the pictures by specifying their number, and in doing so, we recall the files which have logged the results of the 2250 activity, i.e. the number of units and the material used. We can erase the offending portions and replace them with new materials thus modifying the cost and we can modify the plan shape by erasing symbols and re-arranging them.

Questions answered by B. T. Torson, Rolls-Royce (See Page 161)
F. M. Larkin (UKAEA Culham Laboratory)
Q. As an installation which is actually using computer graphics to make money, how much effort on the part of Rolls-Royce has been involved since you have taken up computer graphics, in programming and computing?

A. We installed a 2250 in July 1967. Myself and three other people started work in January 1967 six months before we actually installed the device. I had been at M.I.T. some years before that and at Rolls-Royce we had been preparing for this by discussion with management and so on, to make sure that resources would be available. Since then the team has been building up and is now twelve strong; human effort expended is greater than ten man-years. As regards computer time, we normally operate in a multiprogramming mode and for our applications we use of the order of 5–10% of C.P.U. time.

I would make two more points; entering the field at the time we did, the full time-support of a systems programmer was essential. Secondly, none of the programs which I mentioned contain any complex form of data structure.

C. T. Roberts (Westland Helicopters Ltd.)
Q. Is it likely that full scale lofting will be replaced by computer graphics on a large scale?
A. I don't know about replacing manual lofting. I can certainly see it being used as an additional facility by loftsmen. It is being used in this way by certain companies in the States right now.

E. G. Chadwick (Hawker Siddeley Dynamics Ltd.)
Q. What are the relative costs of doing a job using computer graphics compared to more conventional means?
A. That greatly depends on the equipment you are using. So far the user management has indicated what their urgent problems are and we write the programs to help them solve those. In return they tell us what it is worth. However, it does depend greatly on the hardware and we are not running economically in the sense that each individual job is done more cheaply by graphics. We are still learning, but very shortly some of the programs should be doing jobs more cheaply than by any other means. One program, not in the mainstream of Rolls-Royce activity, is 2250 ECAP—the Electrical Circuit Analysis Program; it is found a great help in saving a lot of bench work. One obtains much more information from a session at the 2250 than one could obtain by taking longer and actually building and testing a circuit.

A. F. Nightingale (Vickers Ltd., Barrow)
Q. Does one require your very large computing capacity to do the sort of work you have been doing?
A. No. At the moment the main computers in Rolls-Royce are IBM System 360 Model 65's—and we use a 100K byte partition on one of them. If you go into computer graphics you must define your problems, note what computer is available, how fast it is and what other work it is going to be doing. If for instance you are trying to do graphics with 30 teletype terminals interrupting continually, conditions are different to those where there are no teletype terminals, and this affects the kind

of problem you can tackle. Beckman Instruments in America working on a small configuration have done some very good work, and STC describe work using a rather smaller system than ours, a 360/40. The amount of work possible is largely dependent upon the amount of money you spend.

Questions answered by R. A. Chambers, Edinburgh University
(See Page 179)

P. B. Sandford-Johnson (The Rank Organisation)
Q. Do you think you are going to get commercially meaningful results out of your learning task study with man-machine interaction?
A. This is very hard to say. In fact, the results are just a step, as one of many projects, towards the ultimate aim of developing interesting, intelligent, heuristic programs. It is not the sort of thing you can point to as having an immediate commercial application, being very much long-term work.

R. Borger (Brunel University)
Q. Are you primarily concerned with producing learning in computers, or are you trying to simulate human learning in computers?
A. We are interested in making the computers learn, in being able to store and evaluate experience and base each decision on experience. Eventually, we hope to progress into the field of concept formation by enabling the computer to form generalisations or concepts based on its experience. However, it is of great help in these interactive studies just to see how the mind alone works and how human minds in co-operation with computers work. This suggests new avenues of approach in developing ideas and techniques.

W. F. Hill (Joseph Lucas (Elec.) Limited)
Q. We have here a case of great general interest, of man-computer interaction involving the closing of an automatic control loop, the possibility of adaptive control of considerable prospective value in achieving automatic optimisation, and also, of course, the possibility of instability in this interaction. Are the procedures used dependent on the exploring of statistical correlations between the different measured parameters and the actions taken or do you use a different adaptive process?
A. The algorithm is fairly simple minded. It uses no really complicated statistical techniques. It is fully written up in the paper "Boxes: An Experiment in Adaptive Control" in *Machine Intelligence* 2, published by Oliver and Boyd in 1968. Briefly, we divide the state space into a number of areas ("boxes") sufficiently small to consider any point within that area the same as any other. The computer records statistics about the past behaviour of the system resulting from the decisions (left or right) taken in each box, decisions initially being taken at

random, and applies an evaluation procedure which decides on the basis of past experience which decision it should take on the next run.

In fact, it's not too simple minded as it has a certain amount of "research mindedness" which is a factor we introduced to prevent the system settling into a final strategy prematurely. By "research mindedness" I mean that from time to time it will make the worse of two decisions just to check that it has got enough data on the subject and that it really is the worse.

We have methods also of blocking out this "research mindedness" after a suitable length of time if the nature of the task does not vary with time.

Question answered by R. Parslow, Brunel University

D. A. Franklin (Medical Research Council)
Q. Computer graphic terminals have been considered as a means of getting at a patient's record. It seems to me that this is rather one-sided, because the implication is that one calls the record up simply to read it. But generally one takes some decision on what is read and fresh information is generated for return to store. The doctor will look at the patient's notes and say, "Yes, we will do this, for the next four days", and this is fresh information which has to go straight back to the file. I believe that a doctor will not sit down at a keyboard and type this information in; the best he might do is to write it down more or less legibly. Then it has to go back into the system straight away and I do not see how this is going to be done. We have heard about touch wires and so on, but these presuppose that the computer knows the various options the doctor is going to suggest, and he may decide he wants to try something he has just read about in the literature which is not in the system. I have not heard yet of sufficient flexibility in this particular situation.

A. The King's College Hospital project has got just this sort of problem, which we deal with at nearly every meeting. The way we approached it is that a group of doctors (who are, in fact, most enthusiastic), are defining a dictionary so that all users will understand the terms used, and we operate a system by which the people can type additions to their notes on the end of the file. (There is a legal difficulty that, once the thing is on, you must not delete anything from your notes and, therefore, we are going to have some problems because somebody will write something up in error, and then have to write another complete note of amendment later on.)

A large part of the problem is for the systems analysts or, a better term, the information analysts, to do a comprehensive study for each type of user. The doctors, the nurses, and the people who are working in the laboratories who are going to fill in information from tests, all have to be able to add to the file.

Among many other problems, there are tremendous security problems because obviously the clerk who does the admissions must be able to put in the name and address, National Health Service number and other data, but is obviously not allowed to know all the other things that are on the file. Again, with psychiatric patients, for instance, even people with very high priority are not allowed to see the psychiatric records.

We have the problem, of course, of somebody watching the doctor in the ward. He may spot how he requests information and then try to use the system himself, because if he quotes the doctor's code number, priority, or whatever it is, then he obviously has access to whatever the doctor has access to. We are arranging a system whereby such a person will not be able to use it on the ward. We want to make sure also that he cannot go to someone's office and start accessing the information from there. We have systems of messages that have to be put in with every entry, so that we have a check on who is doing it.

The actual difficulty of getting the doctors to type in is minimal: they do not want to write notes and then put them in afterwards.

Dr. F. E. Taylor (The N.C.C.)
Comment. I would like to consider another aspect of this, further to Mr. Parslow's remarks. The problem here is one of interaction between the doctor and the record system. Very often, the interaction time can be long. After a doctor has examined a patient, he does not necessarily need his new records brought back to him at the point of use. What we need is a data capture device which is not necessarily a graphics display or a keyboard, but it could well be a magnetic tape recorder, recording the doctor's diagnosis, etc., as voice signals. This information need not be input at the time when the doctor is examining the patient. It can be input later by clerical staff, so that it can be retrieved perhaps in a few hours' time. (A ray of light on the horizon is the IBM "Shoe-box" system which will input voice information straight into computer record form, but this I do not think is yet viable.)

We have to keep our sense of perspective here: graphics is a very useful tool, but it cannot solve all the problems and I think some other form of data capture device can often be used.

D. A. Franklin (Medical Research Council)
Comment. It is true that off-line records do not need to be treated in this way, but I'm really thinking of decisions on treatment which have to be got into the patient's record as soon as possible and also have to be acted on. It is no use, with decisions on treatment that has to take place every two hours, getting the record twenty-four hours later.

Professor B. Herzog
Comment: That is part of the design problem that has to be solved and it requires great effort in the planning of a scheme. There have been many problems in these situations where the design has not yet

been carried out uniformly successfully, but such considerations are essential. Obviously we must get to the point where this can be handled.

Question answered by R. H. Stotz, Computer Displays Inc.

J. S. Clifton (University College Hospital and Medical School)
Q. Another problem we face is that the alphanumeric display will only handle something like fifty per cent of the patient's notes. After all, our patient record also contains analog information in the form of X-rays, of E.C.G. records and so on, and if you try to digitise these you get back an unsatisfactory answer and you fill up half the biggest computer you care to think about. What work, if any, is going on in the development of graphic displays which enable one not only to have alphanumeric information generated by the computer, but also to display information, such as X-rays, etc., simultaneously?

A. At M.I.T. there is some work in information retrieval called Project INTRACS and they are attempting to put the library at the disposal of users at remote stations. One of the problems they face is the enormous quantity of data to digitise, and part of the problem is that of pictures. As they are doing some work, you might contact them at M.I.T. To date they are looking into techniques for image display that are of a more photographic type. A scan picture is sent across and the entire picture is then developed and displayed. The net effect is that you cannot change it, once it is there, but in their situation, as in yours, it is not important to change it. This is contrasted with an interactive terminal where you have to be able to add to what is already there, and is a different kind of a problem. There are a number of things, mostly in the laboratory now, that look promising, but cheap systems for large-scale data display are still several years away.

PART 3
For the Computer Technologist

Present-day Computer Graphics Research

R. ELLIOT GREEN
Scientific Control Systems Ltd.

Introduction

Computer graphics is already proving itself to be an extremely valuable and exciting field. The previous sections dealing with hardware and software systems and their applications inspire an atmosphere of confident anticipation for the future of graphics. It appears to be generally accepted that the present trend towards decreasing terminal costs and increasing system flexibility will be maintained and even accelerated.

The hardware and software of tomorrow are determined by the research projects of today. It was therefore decided to include in this book a review of some of the major areas of research which were discussed at a two-day informal Specialist Session that followed immediately after the Brunel Symposium in July 1968. Many of the most active and advanced workers in computer graphics research, from USA, France and UK, met together to describe and discuss their current projects.

Although reports covered work ranging from the hardware design of plotters to the mathematical representation of three-dimensional curves and surfaces, the major part of the work and the resulting discussion focussed on two main areas of common interest:

1. Low cost remote display systems.
2. Graphics languages and programming systems.

This chapter is therefore confined to these two topics.

Low Cost Remote Display Systems

The pioneer work in computer graphics has been done with expensive highly interactive displays each with its own dedicated processor. It was generally agreed that the graphic terminal of the future must be cheap and must not rely on a local large-scale processing facility. There was notably less agreement on what compromises should be made in order to achieve this ideal. Is a stand-alone terminal with a small local general purpose processor the answer? Should the terminal processor be linked to a, perhaps distant, large-scale processor? If so, how do you divide the computational labour between backup processor and terminal? How should information be best exchanged between the two? Should one optimise on phone line utilisation at the expense of terminal hardware cost? Should minimising terminal hardware be the primary goal? What level of interaction does one need from the terminal? These are some of the problems inherent in the design of such a terminal.

Several papers were presented which dealt with current projects involving low cost terminals of one type or another. Pinson[1] described the GRAPHIC-2 system now in use at Bell Telephone Laboratories. This is a highly interactive system using DEC 340 displays and PDP 9 processors as satellites. These communicate over the public telephone network with a large-scale GE 645 backup processor. In order to keep the data transfer between the two processors to a minimum, they have devised their own programming language, GRIN-2, and an ingenious parallel processing system. The graphics program is compiled on the backup processor into a pair of object programs, one to be run on each machine. Synchronous operation of these programs allows applications programs to be written in a high level language and run on the backup processor, while the remote terminal retains control via its interactive facilities. This requires a low data rate from backup to satellite (150 baud) and a moderate rate in the other direction (2000 baud) and gives terminal response time measured in fractions of a second.

Linders[2] and Kirstein[3] reported on the state of the Imperial (I.C.) and Institute of Computer Science (I.C.S.) computer link-ups. These both use DEC satellites connected to their backup processors over the public telephone network. I.C. use the Univac 1108 at the National Engineering Laboratory near Glasgow and I.C.S. use the IBM 360–75 at the Rutherford Laboratories, Harwell. They each maintain an attention stack of jobs ready to go to the backup machine to be run under its operating system. I.C. plan to use Newman's[4] Reaction Handler system for interactive programming and are at present writing a re-entrant supervisor to handle communications and a job queue in the 1108.

Kilgour[5] went one stage further in cost reduction by describing a system developed at the University of Edinburgh. This uses a 340 display on an 8K PDP7 with no backup computer, and is being operated profitably on a pipeline layout project for a local shipbuilding firm. They have been able to use this small store by curtailing the display versatility. A special graphics program has been written, which uses lists rather than a ring structure and no display file. The display is under program control at all times and the program hesitates after producing each segment of display code to discover whether there was or was not a light-pen-see. The system has 5 modes of operation, each activated by a light button. The requirement for light-pen tracking has been eliminated by arranging for a grid of 128×128 points to be displayed on the screen. Symbols or characters can be positioned at these points and lines made up from short vectors drawn vertically or horizontally between two adjacent points. Used in conjunction with an incremental plotter, this system allows the operator to modify and replot a pipeline layout diagram in half an hour. This task would otherwise have taken a draughtsman a whole day.

The terminals discussed so far are of a now conventional type requiring a small processor, in these cases costing about £10,000, to maintain a refreshed C.R.T. display. Recourse to backup processor is made in the Bell, I.C. and I.C.S. systems to run complex applications programs. For a comparatively simple task the Edinburgh system gains on a cost/efficiency basis, as by doing away with the display file and by using simple data structure, it has storage available to hold small applications programs. It would not be able to compete with the backup systems on anything but the smallest CAD problems.

Ruben[6] described the VD8/I, a new type of system he has developed at DEC. This uses a storage tube and thus relieves the PDP8 processor of the task of storing the display file. It is designed for stand alone operation with up to 8 time-shared displays. It can also be used with a local backup multi-programmed processor with up to 16 displays. The cost per terminal can in this way be reduced to $8,000 with a full cluster. Efficient use of this system remote from the backup processor would require expensive dedicated wide bandwidth transmission lines.

The cluster mode of operation is ideal for a large research or engineering environment, where many users and a large processing facility are all on the same site. There are, however, a large number of potential graphics users in small companies or in outstations of large companies or establishments, where stand alone remote terminals would be required.

The ARDS (Advanced Remote Display Station) unit designed by Stotz and Cheek[7] at M.I.T. is such a terminal. It uses a Tektronix 611 storage tube and operates with its connection to a backup processor over the standard switched telephone network. The processor sends a display file of commands down the line to the terminal, which incorporates special purpose decoding logic and incremental hardware function generators. It offers a somewhat limited interactive facility with keyboard and joystick input. Stotz commented, that from his experience at M.I.T., 75% of the applications for graphics did not require more sophisticated equipment. It was generally agreed that many of the refinements of today's expensive highly interactive display could profitably be traded for a reduction in cost. Stotz has now formed a company manufacturing and marketing the ARDS unit at a basic price of $12,750.

Ruben commented that in his opinion incremental function generation and drawing techniques were not the best approach. He used 12 fairly cheap analog amplifiers in the VD8/I; these give sufficient linearity with a writing rate of 1 cm per millisecond, which is quite sufficient for use with a storage tube.

Pitteway[8] suggested doing away with both incremental and analog function generators. He felt that the future of graphics depended on reducing the terminal cost so that it was competitive with a teletype. If fully descriptive graphic data could in some way be highly compressed, it could be sent from the processor over a phone line to be displayed on a simple cheap terminal. He described a data compression technique, which he had devised and is being implemented by his group at Brunel. This describes curves by a set of increments somewhat similar in manner to an incremental plotter. This is then coded into a compressed format for transmission and storage. A bit stream, which describes each increment by an average of less than two bits, will then be sent asynchronously to the terminal's sequential store for C.R.T. display refresh.

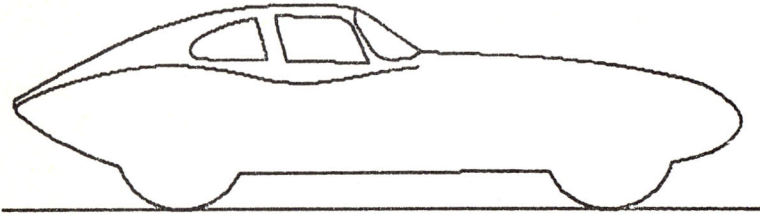

Fig. 1 Automobile silhouette plotted by increments

Figure 1 shows an automobile silhouette, which was plotted involving 1957 drawn increments from a file containing 2470 bits of compressed code. This includes initial pen positioning and works out at 1·26 bits per increment. The compression technique involves splitting up a picture into a set of short discrete curves. The direction and curve length are specified; each subsequent bit then initiates a parallel or a diagonal move. If each curve is given a name, this could be stored in a buffer while the curve is being displayed and could be sent back to the computer as part of an interrupt as a result of a light-pen-see.

Ninke[1] described GLANCE-2, which is a similar system being developed at Bell Telephone Laboratories. It will have stand alone refreshed C.R.T. remote displays, using incremental vector and character generators in conjunction with an efficient data compression technique. This system will also allow a limited amount of interaction with a light pen. Ninke gave an example of how the terminal, without recourse to the processor, could indicate a light-pen-see on an entity by displaying it more brightly than the rest of the picture. He is using MOS integrated circuit dynamic shift registers as buffer stores and predicted that their cost, over the next six years, would drop from 15 cents per bit to 0·1 cent per bit. He plans to use read only memories as character font stores for a local character generator.

Both the Bell and the Brunel systems are, in varying degree, simplifying the terminal hardware in an attempt to reduce its price to an acceptable level. This, in spite of data compression, is a trade off against phone line utilisation. A more sophisticated terminal such as ARDS with integral hardware function generators requires lower data rates from the processor to describe its picture. There was much discussion and little general agreement as to which, if either, was the optimum compromise. There was also discussion about the relative advantages and disadvantages of storage tubes and refreshed C.R.T.s as remote display devices. The main advantages of the storage tube are that it requires simpler and hence cheaper circuity to operate and that no buffer store is required. Tube life is, however, limited and tube replacement costs were estimated at about 40 cents per hour. The refreshed C.R.T. has the advantages of larger tube size, greater brightness and the very valuable ability to pick with a light-pen. Also if a part of a picture has to be changed, the whole display file does not need to be retransmitted. The participants were divided as to which, if either, they thought would be best incorporated in the ubiquitous remote terminal of tomorrow.

An entirely different hardware approach at Reading University was reported by Lee.[9] They are experimenting with a standard TV set as a display device with a sequentially scanned Williams tube as its store. They have also produced an extremely cheap graphical input device using Teledeltos resistive paper as a 2D drawing pad. One of the major problems with the use of a TV raster scan display is that it requires a large amount of store. This will be supplied quite cheaply by the Williams Tube. It is unlikely to be of much value as a remote terminal, however, as the data rate required to fill the store would be very high.

Graphics Languages and Programming Systems

The discussion of graphics languages and programming systems was far more fragmented and less specific. There was a common plea for a universal graphics language for computer aided design, but no definitions for the formulation of such a language were forthcoming.

For CAD use, a graphic terminal requires software backup at differing levels of specialisation. These can be visualised as three main levels, each laid one on top of the other. The lowest interacts directly with the hardware and handles such functions as interrupts, autonomous peripheral transfers, hardware function generators and display file maintenance. This level would be written in the assembly language of the dedicated processor and is usually supplied by the terminal manufacturer.

The second level incorporates data structure and controls the formulation and maintenance of a mathematical model of the object being displayed. It must also supply graphic routines, which are used at the upper level and which control functions in the lowest level. Most terminals being used industrially for CAD use the manufacturer's software at this level. Parts of this would probably have been written in assembly language and other parts in a higher level language.

The top level comprises applications programs. These are designed to solve particular problems using the graphic terminal. These would normally be written in an extended version of a conventional high level language, and would not be supplied by the terminal manufacturer.

The engineer or scientist who uses the terminal merely controls all these levels of program, and does not need to do any programming himself.

The graphics language, which was felt to be so desirable would be for use by applications programmers. At present most applications programs are written in FORTRAN and utilise a language of graphic calls, which are, in fact, a set of FORTRAN routines embedded in the next lower software level. It was generally agreed that FORTRAN, or in fact any high-level procedure oriented language at present available, was very clumsy and unsuitable for interactive graphics. This is primarily because of an inherent inability to handle real time interrupts and the lack of built in graphic i/o commands.

FORTRAN is however being used by most active CAD groups because it is all that is available. Torson[10] from Rolls Royce stated that until something better was available graphics experience must be gained using present-day languages. He suggested that existing non-graphic applications programs should be modified in gradual stages to take advantage of the display. He emphasised that an existing relevant program which had been modified is worth any number of clever and sophisticated programs, if they are not relevant to the job in hand. Also the most urgent requirement in the CAD software field was for packages, which can be used in existing high level languages, on commercially available machines—e.g. data structure packages incorporating model segmentation and 3D geometry.

Wycherley[11] and Rundle[12] outlined the approach they had taken at Elliotts (now ICL). The lower two programming levels are handled by a package called DISMAN. This controls:

(i) all conversion of non-graphic data into graphic commands for the display,
(ii) a data structure and organisation and control of a display file made up from this structure,
(iii) the first stages of all on-line imputs including light-pen-sees. They have written two interface packages, which allow applications programs written in ALGOL or FORTRAN to interact with DISMAN via subroutine calls.

They have also recently implemented a version of the ASP ring structure package, which can be utilised in FORTRAN. This allows the applications programmer to get at the data structure and gives him considerable flexibility in defining the way in which his model is structured.

Rafferty[13] from the National Engineering Laboratory and Larkin[14] from Culham Laboratory, both presented packages, which allow graphic output from FORTRAN applications programs. The Culham system, GHOST, is built up from a language containing a large selection of FORTRAN subroutines. These include graphic procedures and library routines for operations such as stereo projection and histogram plotting. Special routines control the mapping from mathematical (M) space as in the data structure to plotter (P) space. Data can be fed to C.R.T. terminal or plotter, or both simultaneously with automatic scaling to suit their appropriate P space. Data can be stored in terms of M space, e.g. 3-dimensional coordinates relative to the origin of the model's coordinate system or in P space, e.g. 2-dimensional coordinates relative to the bottom left-hand corner of the plotter table.

Bird[15] discussed the Marconi approach to graphics software. Operator inputs generate ring structure containing blocks for drawings, display elements, points, values and constraints. He stressed the importance of integrating the problem data structure with the model data structure. A FORTRAN applications program can be used by the operator calling it, and then listing on the screen all the names he had used in the diagram (i.e. in the model) in the order required by the program.

There was much involved discussion of data structure, with particular reference to parallel processing for remote operation. Ross[16] made the descriptive comment that parallel processors were like two mice running around the data structure, each leaving messages for the other. He reported on the state of his AED project at M.I.T. AED (Automated Engineering Design) is a very comprehensive ring structured extension of ALGOL for use on large time-shared computers. The AED compiler is at present being bootstrapped to several machines; so far this has been completed on the Univac 1108 and IBM system 360.

Newman[4] commented that the systems discussed so far are designed primarily to provide graphic output. Meaningful graphic input was a far greater problem and had been tackled by very few researchers. He discussed his "Reaction Handler" system, which he originally devised at Imperial College and was now working on at the University of Utah. This allows one to write and debug interactive programs graphically on the display. State diagrams defined in a Network Definition Language are used to describe graphical operations of an interactive program. This program has two separate components (i) functional routines written in a high level language and (ii) control routines which are defined in terms of a set of program states. The control component is designed on the display by making appropriate connections between various states and by calling in functional routines. The two components of the programs are compiled separately and are run together under a supervisor called the "Reaction Handler". Figure 2 shows a state diagram defining a rubber-band line drawing program. Before entering state 2 "storpoint" a high level procedure is called to store the starting point of the line; thereafter each time the light pen is moved the procedure "drawline" will compute and display a fresh line. Also shown in Figure 2 are a list of commands chosen from light buttons, which actually call the various components of the diagram on to the display.

Several of these state diagrams can be put together to form a fairly complex interactive graphic input program.

Lang[17] reported on a new project at Cambridge University to represent and manipulate 3-dimensional objects on the display. A graphic terminal backed by storage and computation has the potential to display a 3D object far more realistically than the conventional orthogonal projection drawings. Facilities offered by the terminal include rotation with hidden line removal, production of cross sections, showing moving parts in motion and production of stereo pairs. One of the major problems at the moment is a lack of methods of defining 3D objects so the computer can store and display them in such a way that they can be subsequently manipulated.

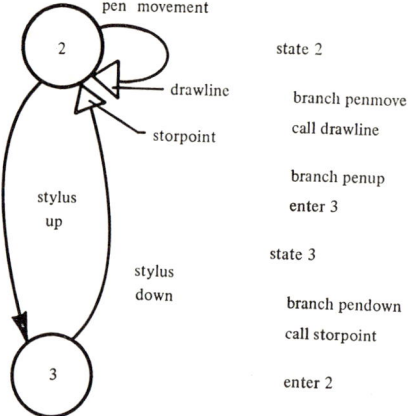

Fig. 2 State diagram for rubber-band line and light button commands

At Cambridge they have defined basic 3D objects such as parallelepipeds, cylinders, etc. Positive and negative volume are used, thus a +ve. cube and a −ve. cylinder could represent a cube with a hole through it. These can be isotopologically transformed, i.e. rotated, translated, scaled overall or independently in the x, y or z directions, and fitted together to form more complex volumes. Figure 3 shows the early stages of an attempt to design a sofa from these basic objects. It includes a volume which has been called up and has yet to be transformed and added to the sofa as a second arm.

An ASP data structure is used to model the various volumes and assemblies. This contains rings and associators so individual volumes can be manipulated as discrete entities or can be amalgamated to form discrete assemblies. Newman's Reaction Handler system was used to implement the language with which the designer builds up the assembly.

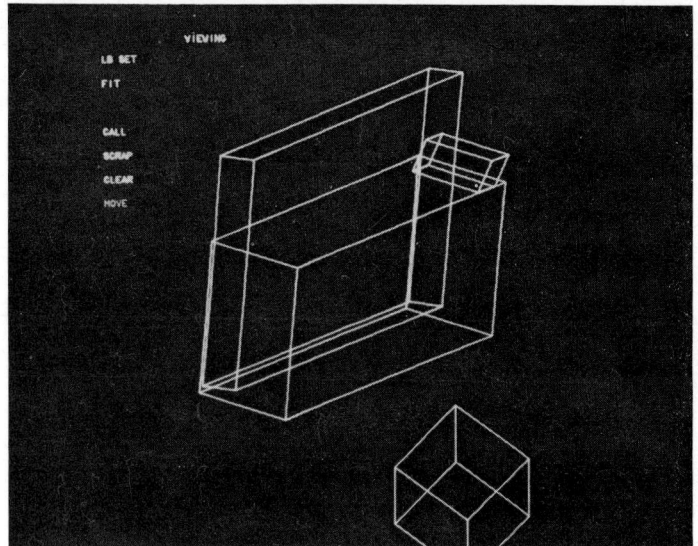

Fig. 3. Attempt at sofa design from 3D objects

Conclusion

Research effort in computer graphics is going into many other fields. However, the emphasis is on reduction of terminal cost, widening of system flexibility and easing of operating complexity. The fruits of this research will be the continuing increase in the application of computer graphics to an expanding range of environments.

Biographical Note–see page 107.)

Contributors

1. Pinson, E. N., and Ninke, W. H.   Bell Telephone Laboratories, U.S.A.
2. Linders, J. G.   Imperial College, U.K.
3. Kirstein, P. T.   Institute of Computer Science, U.K.
4. Newman, W. M.   University of Utah, U.S.A.
5. Kilgour, A. C.   University of Edinburgh, U.K.
6. Ruben, M. A.   Digital Equipment Corp., U.S.A.

7. Stotz, R. H., and Cheek, T. B. Computer Displays Inc., U.S.A.
8. Pitteway, M. L. V. Brunel University, U.K.
9. Lee, B. J. Reading University, U.K.
10. Torson, B. T. Rolls-Royce Ltd., U.K.
11. Wycherley, R. D. H. I.C.L., U.K.
12. Rundle, A. R. E-A Systems Ltd., U.K.
13. Rafferty, J. S. National Engineering Laboratory, U.K.
14. Larkin, F. M. U.K.A.E.A., Culham Laboratory.
15. Bird, S. The Marconi Company, U.K.
16. Ross, D. T. Massachusetts Institute of Technology, U.S.A.
17. Lang, C. A. Cambridge University, U.K.

PART 4
For Reference

Some Commercially Available Computer Graphics Systems

A major feature of the Brunel Symposium was an exhibition of on-line equipment operating either from a local computer or over telephone lines to a remote backup processor. Manufacturers also presented lecture-demonstrations on applications of their systems, and participants were able to see and operate the graphic equipment.

The exhibition complemented the main session by allowing delegates to become acquainted with existing equipment, and it was felt that readers of this book would welcome the opportunity to have similar information on commercially available graphics systems. Manufacturers were therefore invited to submit advertising material on their products and this appears on the following pages.

DIGITAL GRAPHIC SYSTEMS

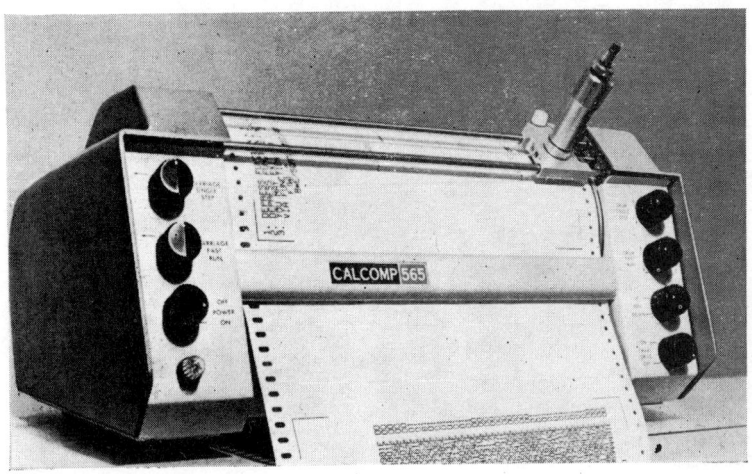

CALCOMP Digital Plotters, with on-line Controllers or off-line Magnetic Tape Units to drive them, provide graphic output facilities for all types of computer. Among the many machines which can now turn numeric figures into meaningful graphs and displays by use of a complete graphic system, are those of BURROUGHS, C.D.C., D.E.C., ELLIOTT, ENGLISH ELECTRIC, FERRANTI, G.E., G.E.C., HONEYWELL, I.B.M., I.C.T., R.C.A., S.D.S. and UNIVAC. By a complete graphic system, we mean a system selected from a wide range of Drum, Flatbed, and Electronic Plotters, supplied together with special chart papers, pens and accessories, PLUS—**FULLY OPERATIONAL SOFTWARE PROGRAMS.**

DRUM PLOTTERS	CalComp drum plotters are available in two sizes, 12 in. and 30 in. They use special chart paper and can produce plots up to 120 ft. in length.
FLATBED PLOTTERS	Also available in two sizes, 31 × 34 in. and 54 × 72 in. These provide continuous display during plotting and can draw on to preprinted forms or special materials.
ELECTRONIC PLOTTER	The CalComp model 835 plotter generates its plot electronically on the face of a CRT and this is automatically recreated on microfilm. This allows very high plotting speeds.
ON-LINE SYSTEMS	CalComp offers a wide selection of interface units for on-line digital plotting with most standard computers, including the new time shared systems.
OFF-LINE SYSTEMS	CalComp magnetic tape systems provide the special features required for low speed reading of standard format tape for off line plotting.
SOFTWARE PROGRAMS	Basic Programs and applications packages tailored for your computer configuration and operating systems provide all the software needed to obtain general purpose graphic output from the plotters.

CALCOMP LIMITED
BURGOINE HOUSE · HAMPTON WICK · KINGSTON · SURREY
Telephone: 01-977 7612

6000 Series Incremental Plotters

Highspeed automatic drawing

30″ wide plotting paper

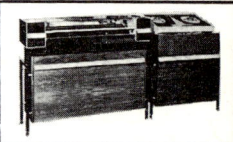
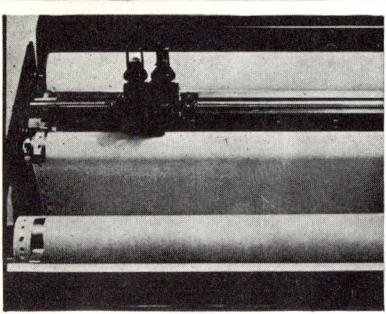
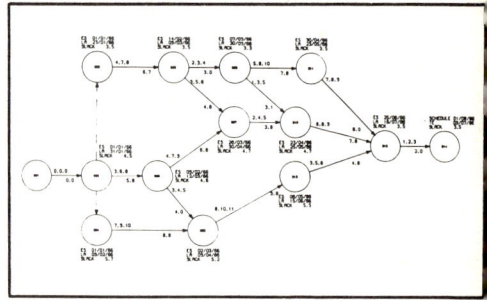

On-line operation with most digital computers, or off-line systems with punched paper or magnetic tape output.

6000 Series Incremental Plotters produce fully annotated drawings and plots either on-line with most digital computers, or off-line from magnetic tape or punched paper tape. The 6000 Series will accept paper 30″ wide (in 120′ rolls) and operate at a speed of six hundred ·005″ increments per second. Options include two-pens carriage, each pen selectable under input control; an adaptor to accept 14″ wide paper, and a larger increment size of ·010 with a speed of 300 increments per second.

For further details telephone or write to:

Datagraphic & Digital systems
Computer Instrumentation

Computer Instrumentation Ltd., School Lane, Chandlers Ford, Eastleigh, SO5 3YY, Hants, England
Telephone: Chandlers Ford (STD 0421 5) 4671   Telex: 47326

COMPLETE SYSTEM CAPABILITY

Over 7000 DIDS 400 are in use and on order, and are NOW performing these functions:–

- Inventory Control, Scheduling and Production Systems
- Airline Reservations
- Medical Records
- Message Switching
- Air Traffic Control
- Customer Service Enquiry Records

Cossor Electronics Limited have design teams specialising in visual display systems supported by full production, installation, service and training facilities at Harlow, Essex.

Illustrated is a standard DIDS viewing unit and a DIDS. 421 control unit in final test. The full range of DIDS. 400 equipment is now in quantity production.

COSSOR ELECTRONICS LIMITED   DATA SYSTEMS DIVISION
The Pinnacles, Elizabeth Way, Harlow, Essex. Tel. Harlow 26862. Cables Cossor Harlow Telex 81228

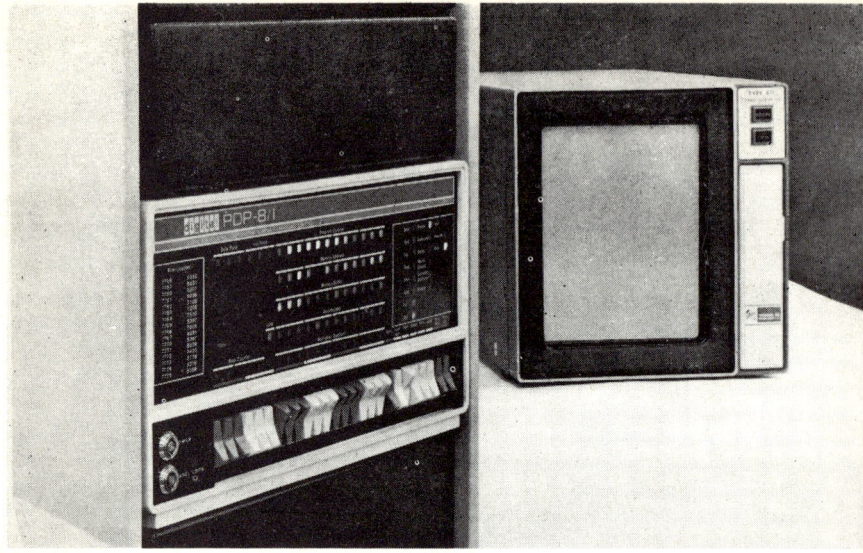

New Low Cost Graphics Capability For Digital's PDP-8 Series Computers

DIGITAL'S new low cost graphics terminal with KV8/I controller and EDGRIN software package provides display capabilities found only on larger, more expensive graphics systems.

Features include:
- Linear Stroke Vector Generation.
- No Refresh necessary.
- Large Viewing Area (21 × 16·3 cm.).
- On-Line Editing for Picture Manipulation.
- Variable Font Character Generator.
- Iterative Sub-Routines for Grid Mapping.
- Joystick Pointing available for Linking Sub-Routines for Display File Construction.
- Hard Copy and/or Paper Tape Output.
- Disk provides additional buffer capacity to allow expansion of executive programmes in core.
- Complete with general purpose graphics software package ready for immediate use and easily expandable by the user.
- Generates MACRO programmes that can be referenced as sub-routines, eliminating the necessity of a detailed description of each sub-picture during multiple usage.
- Light Buttons for Programme Interaction.
- Single Command Curvilinear Circle Generation.
- Text Editor permits On-Line changing of display copy.

EDGRIN software provides an interpreter which translates graphic information into machine language; an editor which allows easy debugging of graphics programmes; and an assembler which creates a paper tape compatible with other PDP-8/I assemblers. The system also comes with a customer proven PDP-8/I software package which includes FOCAL, Digital's powerful new conversational computer language.

For further information on the new graphics system contact your local DIGITAL sales office, or write to Digital Equipment Co. Limited, Arkwright Road, Reading, Berkshire.

Computers · Modules

Digital Equipment Co. Limited, Arkwright Road, Reading, Berks., Telephone: (ORE 4) 85131; 13/15 Upper Precinct, Walkden, Worsley, Manchester, Telephone: (061) 790 4591. Offices also in Paris, France; Munich and Cologne, Germany; Stockholm, Sweden; The Hague, Netherlands; Oslo, Norway; Sydney and West Perth, Australia; Carleton Place, Toronto and Edmonton, Alberta, Canada; And principal cities of the U.S.

 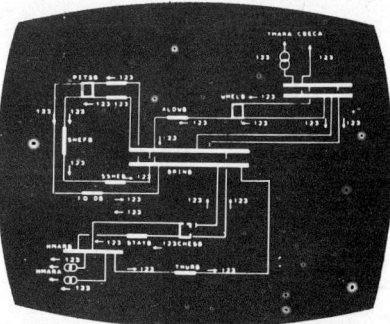

ARGUS FLEXIBLE DISPLAYS

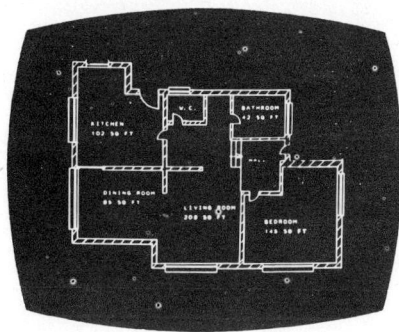 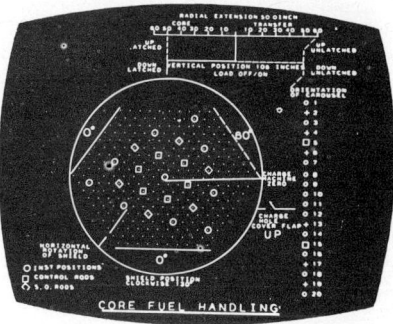

To meet the need for a simple, reliable and effective man/machine communication facility in computer controlled process systems, Ferranti have developed the ARGUS range of real-time CRT displays. ARGUS displays offer outstanding and unique flexibility of operation; the Display Keyboard can be used to input information to the computer store, to extract information for visual display, for the editing of alpha-numeric data, and also provides drawing facilities for symbols and diagrams. Software facilities provide for six modes; the writing of random characters, tabular data and symbols, and the drawing of lines, circles and graphs, which can have fixed or random increments.

ARGUS Displays are of modular construction, and can be tailored to individual requirements using standard modules. They are suitable for all ARGUS Computer Systems, or, when provided with custom-made interface, with any other digital computer system.

The flexibility of ARGUS Display Systems has led to a wide variety of applications, including a number in the Nuclear field, Gas and Electricity distribution, and in B.O.A.C.'s International Seat Reservation System BOADICEA, in which seven hundred ARGUS Displays are used associated with thirty ARGUS 400 Microminiature Digital Computers.

Full details, technical literature etc. available from FERRANTI LTD., AUTOMATION SYSTEMS DIVISION, SIMONSWAY, WYTHENSHAWE, MANCHESTER M22 5LA Telephone: 061-428 3644.

FERRANTI
ARGUS 400/500 SYSTEMS

FAS 30

C.R.T. DISPLAYS— 'ETOM' RANGE 2000 - 5000

▲ Allows high speed data transfer of pictorial information.
▲ Data stored locally for screen re-fresh.
▲ Large 20″ screen.
▲ Ability to digitise original sketches.
▲ Remote or local working.
▲ Low cost.

 AN ELECTRAUTOM COMPANY

GRAPHIC DISPLAYS LTD

38A CROMWELL ROAD,
LUTON, BEDS.

Telephone: 0582 30847/8
Telex: 263498 A/B ETOM LDN

HEAD OFFICE: 408 FINCHLEY ROAD, LONDON N.W.2

BUFFERED LINE INCREMENTAL MAGTAPE PLOTTING—'BLIMP'

▲ Improves Computer Plotting Efficiency.
▲ Stores up to 250,000 Plotter moves.
▲ Allows 25 Kc/s Transfer.
▲ Off line playback at 300–1000 inc/sec.
▲ Compatible with all incremental plotter formats.
▲ Low cost.

GD 4000 MAGNETIC STORAGE DRUMS

▲ 4096 × 8, 16, 2 × 32 bit words.
▲ Transfer 200 Kc/s.
▲ Maximum access time 20 ms.

DIGITAL INCREMENTAL PLOTTERS—'COMPLOT'

▲ 300 increments per second at 0·005″, 0·010″, and 0·1 mm and 0.25 mm.
▲ 11″ or 22″ flat bed and fan fold paper.
▲ 3 alternative pens supplied.
▲ Extensive soft-ware available.
▲ Interfaces for on line or telephone line working.
▲ Magnetic or paper tape off line systems.

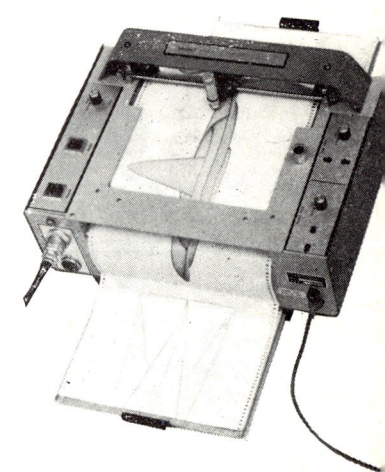

IBM demonstration of graphics in engineering and science

This IBM 1130/2250 Graphic Data Processing System was shown at Brunel University in July. Visitors were able to experiment with the light pen, using programs available from the system's Disk Storage.

The 1130/2250 is a powerful and direct method of man-machine communication. It gives designers, engineers and research scientists instant visual access to results of their computer programs. And the information presented is in an immediately usable form. Using the IBM Graphics Subroutine Package, you can easily write your own programs. And with the latest Disk Monitor Operating System, now available, the full power and flexibility of this disk based graphics system is at your command.

The 1130/2250 can also be linked to IBM System/360 thus giving access to the facilities of Operating System/360. If you would like a leaflet and more information about the IBM 1130/2250, write to Mr M. J. Evans, IBM United Kingdom Limited, 389 Chiswick High Road, London W4.

IBM

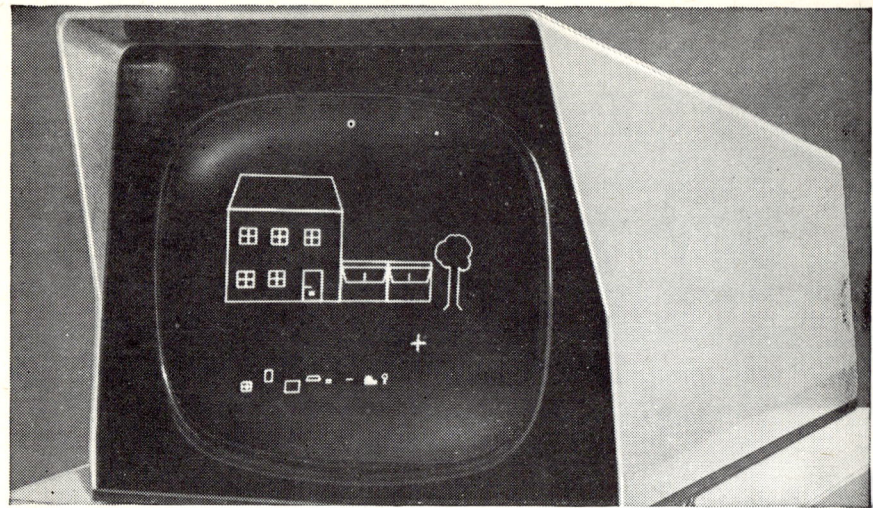

ICL gets you home with the latest in computer graphics

If you alter a design or engineering calculation, the ICL 4100 interactive graphic display system shows you instantly the effect of your alteration. If your drawing contains standard components, like our house above, these can be stored as sub-pictures and placed in position as required. The benefits of these and other facilities are substantial. As an independent evaluation* bears out, the 4100 represents, by a powerful combination of hardware and software features, advanced display technology at low cost.
*Computer Display Review published by Adams Associates.

Demonstration and display
The material used in our demonstration at Brunel including slides and film is now available, together with current display material describing applications. For full details write to John Rodgers at 113 Upper Richmond Road, Putney, London, S.W.15.

The Computer Industry

ICL
International Computers Limited

Marconi graphics

X2000 DATA DISPLAYS AND MYRIAD REAL-TIME COMPUTERS

Powerful man-machine interactive systems

Comprehensive input facilities
standard computer inputs plus light-pen, touch-wires, tracker ball and special purpose keyboards.

Modular system
single control unit drives up to six displays each with independent format and operation. Optional buffer storage, with one or two control units, hardware or software character generation, disc and magnetic tape backing stores.

Versatile graphic facilities
plain broken, thick or thin lines can be drawn between any points on 1024x1024 matrix. Symbols and sub-pictures by sub-routines.

High definition
new advanced technique in c.r.t focusing results in exceptional picture clarity.

Graphic software
based on an open-ended ring structure technique providing a powerful base for application-orientated graphic systems.

Satellite operation
system can operate as a satellite to a large data processor over an ordinary telephone line where extensive computation is required.

Hard copy outputs
permanent records readily obtainable on paper or film.

Standard software
including Algol, Fortran, Coral, Myriad User Code and Operating Systems.

THE QUEEN'S AWARD TO INDUSTRY
1966 1967 1968

Marconi Myriad computers and X2000 data displays

The Marconi Company Limited,
Automation Division,
Marconi House,
Chelmsford, Essex, England
Telephone: Chelmsford 53255
Telex: 99201

AN ENGLISH ELECTRIC COMPANY

LTD/AX52

SERIES/F ELECTRON BEAM RECORDING SYSTEM

Direct conversion of computer tape output to microfilm images

Output dissemination speeded
Disseminated information more easily updated
Computer time and storage capacity saved
Reference to computer output quicker and simpler
Storage problems solved

The Electron Beam "Writes" a Latent Image on Dry-silver Microfilm.

Computer Micro-Graphics

The 3M Series F. Electron Beam recording system enables computer generated digital information stored on magnetic tape to be directly converted into readable microfilm print-out at speeds compatible with tape units – up to 30,000 lines per minute. Forms merging is also possible, ranging from simple tabulations rules to exploded-view diagrams.

Applications

One of the most important benefits of Series F. EBR computer Micro-graphics is its use for creation of a microfilm memory bank, relieving computer capacity. In a typical public utility installation the equipment is used for creating microfilm images of customer transactions from magnetic tape. Reference to $\frac{1}{2}$ million current accounts is made on demand from the computer, while back reference up to several months is made from cartridged 16mm. microfilm, cross-referenced on the computer terminal display. With some 300 characters in memory, this computer has a mass storage of 150 million characters memory, while ten times this amount is stored on microfilm, relieving the computer capacity and obviating expensive programming and storage. Many other applications are apparent where a large volume of computer-produced documentation must be referred and stored, such as banks, insurance companies, retail stores, government organisations and universities.

Microfilm Products
3M HOUSE · WIGMORE STREET · LONDON W1A 1ET
TELEPHONE (01) 486 5522

FLICKER–FREE CRT STORAGE DISPLAY

The new Tektronix Type 611 is a complete console and remote terminal readout device presenting alphanumeric and graphic information to very much higher standards of clarity than hitherto available. The 11-inch magnetically deflected, bistable storage display tube offers:

- High information density—4,000 characters 90 × 7 mils in size may be clearly displayed with good spacing on the 21 cm × 16·3 cm screen.
- Excellent resolution equivalent to 400 stored line pairs along the vertical axis and 300 stored line pairs along the horizontal axis.
- Dot setting time is 3·5μS/cm + 5μS and dot writing time is 20μS.
- Time required to erase and return to ready to write status is 0·5 seconds.
- Operating functions are remotely programmable through a rear panel connector.
- Accepts X, Y and Z inputs from peripheral equipment.
- A "Write-through" feature provides an index to the writing beam position without storing new information or altering previously stored information.

Type 611 Storage Display Unit .... Price £1183 · Duty £194. 8s.
Other units in this range include:
Type 601 5-inch Storage Display Unit   Price £496 · Duty £81. 10s.
Type 602 Non-Storage Display Unit   Price £320 · Duty £52. 14s.

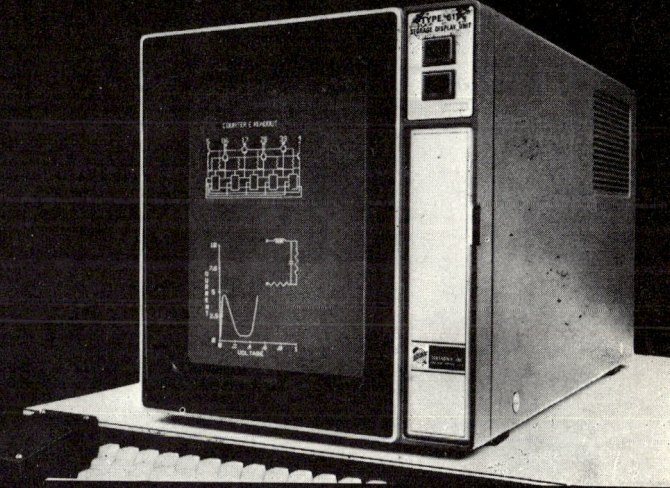

TEKTRONIX U.K. LTD
BEAVERTON HOUSE · HARPENDEN · HERTS
TEL: HARPENDEN 61251 · TELEX: 25559

For overseas enquiries: AUSTRALIA: Tektronix Australia Pty. Ltd., 4/14 Foster Street, Sydney. CANADA: Tektronix Canada Ltd., Montreal, Toronto & Vancouver. FRANCE: Relations Techniques Intercontinentales, S.A., 134 Avenue de Malakoff, Paris XVI. SWITZERLAND: Tektronix International A.G., P.O. Box 57, Zug, Switzerland. REST OF EUROPE AND THE MIDDLE EAST: Tektronix Ltd., P.O. Box · 36, St. Peter Port, Guernsey, C.I. All other territories: Tektronix Inc., P.O. Box 500, Beaverton, Oregon, U.S.A.

UNIVAC UNISCOPE 300
Advanced CRT Display Terminal

From ten years experience with operational real-time systems, UNIVAC has designed the UNISCOPE 300 to incorporate maximum ease and flexibility of input, fastest response, and reliability of operation, (including fail-soft options for the critical environment).

UNISCOPE features include :
- ☐ Data Display of 512 characters (8 x 64) or 1,024 characters (16 x 64)
- ☐ Ergonomic keyboard with standard key-set plus :—
    - Special purpose "Function" keys
    - Variable "fixed data" insert cards
- ☐ Automatic data editing functions
- ☐ Automatic data compression for optimum use of transmission lines and fastest response
- ☐ Single or multi-station control units with up to 48 UNISCOPES per control
- ☐ Full duplex transmission on lines of between 2,400 and 40,800 baud using standard ASC II code.
- ☐ Fail-soft protection including :— Dual control unit option
    Dual transmission—line option
    (both of which contribute to the system performance normally).
- ☐ Long term reliability and durability.

UNIVAC Division Sperry Rand Limited
65 Holborn Viaduct, London, E.C.1. Telephone : 01-236 1010

Glossary of Computer Graphics Terms

This glossary is intended to help the non-specialist reader by giving a brief idea of the meaning of any terms used in this book which might be unfamiliar to him. It is not an exhaustive list of computer graphics terms.

Access bus
    A common set of signal lines to allow transfer of data into or from the store from many different devices.
Alphanumerics
    Letters and numbers as opposed to drawings.
Analog(ue)
    Representation of a variable by a physical quantity, e.g. a voltage, proportional to that variable.
ASCII
    American Standard Code for Information Interchange. (See page 192.)
Assembler (Assembly) language
    A low level programming system which is more convenient to use than the computer's own machine code.
Associator
    Mechanism in a data structure for associating entities.
Attention stack
    A queue of requests for processing by a remote time-shared terminal.
Autonomous access
Autonomous data transfer
Autonomous peripheral transfer
    A system that allows transfer of data into store from peripherals without stopping the computer in its other work.
Bandwidth
    The range of signal frequencies to which a device can respond. For transmission of data, the greater this range, the greater the possible data rate. (See Modem.)
Batch processing
    Processing is carried out on one or more tasks, which when completed are replaced by another batch.
Baud
    A transfer rate of one bit per second. (See page 100.)
Bit: A BInary digiT
    The basic unit of information in a computer. May have the values 1 or 0. Four such bits are required to represent a two digit decimal number up to 15.
Blink
    A flashing item on a display to attract attention. (See page 55.)
Block   A discrete set of data held in store.
Bootstrap
    A program or device designed to bring itself into a desired state by its own action.
Bridge circuit
    An electrically balanced circuit. Any imbalance due to operation of touch wires is detected.

Buffer
: A temporary store for information.

Button box
: An auxiliary keyboard usually with about half a dozen keys or buttons.

Byte
: A group of 8 bits, commonly used as the smallest independent block of data transferred through a system. (See Bit.)

C.A.D.
: Computer Aided Design.

Category Selection
: Selection of data for display. (See page 45.)

Character font
: The basic character shape.

Character generation
: Production of electrical signals to make the C.R.T. beam move over the screen tracing the shape of a character.

Commonality
: The degree of similarity between displays used for different applications (See page 54.)

Compiler
: A program that will translate a high level language into the basic codes usable by a computer. (c.f. Assembler.)

Core store
: The main store or memory of a computer, consisting of many small ferrite rings capable of storing "bits" of information. Groups of bits form significant blocks of data, and may be retrieved and processed. See "Bit".

C.P.U.
: Central Processing Unit, as opposed to the whole computing system.

Cursor
: A moveable symbol drawn on the screen.

Cycle Stealing
: A method used for autonomous access (q.v.)

Dark Trace Display
: A new development of the DVST. (See page 13.)

Data Structure
: The overall organisation of picture data. (See page 19.)

De-bugging
: Removal of "bugs" from a program to make it work.

Dedicated system
: A system where the computer does nothing else but service the display.

Delay line
: A storage system where data is fed into one end of the line, taken out at the other end some time later, and recirculated if still to be stored.

Digital
: Representation of a quantity in numeric form, e.g. decimal or binary numbers.

Digitise
: To transform a diagram into numeric form suitable for processing by a computer.

Glossary

Direct View Storage Tube (DVST)
　A C.R.T. capable of holding a picture without refresh. (See page 13.)
Display File
　The stored data used by the display to draw a picture. (See page 17.)
Dynamic Shift Register
　A circuit which inputs, stores and outputs a stream of data bits.
Entity
　A block of data.
Executive program
　The set of programs to control the running of the computer.
Flying spot digitiser
　A system that encodes a picture by digitising each point depending upon its brightness. (See Digitise.)
Formatting
　Writing of data at fixed positions on a display. (See page 53.)
Function Generator
　Hardware for generating functions such as circles on the display.
Function Keys
　A set of keys for controlling the display program. (See page 10.)
GO TO statements
　Programming statements which alter the sequence of operations in a program.
Gravity Fields
　A method for rapid positioning of lines. (See page 34.)
Hardware
　The actual machinery of a computing system. (cf. software.)
Hardwired
　Hardwired facilities are those included by wiring in of circuits, i.e. in hardware.
Hash coding
　A method of coding data for efficient storage.
Hidden line removal
　Removal of lines describing a 3D object, when they would normally be obscured from view by the body of the object.
High level language
　A programming system where simple statements written by the programmer may accomplish a great deal of computation, through interpretation by a compiler into the more complex programs used by the machine. (See Compiler.)
Holograms
　A system whereby optical interference patterns can be formed by photographing a direct source of coherent light and its rays reflected from an object. Viewing these patterns (holograms) under suitable illumination gives a 3D image of the object. Computer generated holograms might give 3D displays.
Hybrid computers
　Computers using sections of digital and analog circuits in the one machine.
Incremental dots and characters
　Drawings or text built up by displaying adjacent dots or characters in

turn. Little time is wasted moving from one to the next. (Cf. Random dots and characters.)

Integrated circuits
Complete circuits constructed on a chip of silicon without therefore connecting in separate components. The whole circuit generally very small and very fast.

I/O
Input/Output. Used to describe the systems for transferring data to or from the computer.

Interactive Displays
Displays where the operator can modify the display with a light pen or similar device.

Interrupt
A signal sent back to the computer from a peripheral, which interrupts the computer's operation so that it can be actioned.

Joystick
Similar in principle to a rolling ball, but is a lever pivoted at one end. As the other end is moved to and fro, pick-offs detect the movement and move the cursor on the screen. (See Rolling ball, Cursor.)

K
Thousand. A 4K core store contains about four thousand words of storage.

Light button
A symbol on the screen which can initiate a certain program action if the light pen is pointed at it. Similar in operation to pressing a button on a button-box.

Light pen
A light sensitive device used to control a display by pointing it at C.R.T. screen. (See page 11.)

Light-pen-detect, -interrupt, -hit, -see
The detection by the light pen of the phosphor flash, as the computer draws on the display.

Light-pen-pick
To indicate a displayed object to the computer by pointing at it with the light pen.

Lissajous figure
A pattern produced on a display. (See page 10.)

Logic
The basic computer circuitry. Being designed on principles of algebraic logic it is generally referred to as "logic".

Menu
A set of symbols on a display allowing operator to select next action. (See page 32.)

Meta-language
A language which describes the syntax of a programming language.

Mode bits
One or two bits in a word of data specifying, by their value, how that word is to be interpreted. (See Bit, Word.)

Model Segmentation
Splitting up of the mathematical model of the object to be displayed so that parts of it can be in core and other parts in backing store.

Glossary

Modem
Modulator/Demodulator unit used in data transmission. (See page 100.)
Modular Memory
The use of several blocks of memory in a computer. One block may be used for processing, and another for refreshing a display. The processor only interferes with the latter process for updating.
Monoscope
A special cathode ray tube system. (See page 9.)
M.O.S.
A type of circuit element using Metal-Oxide-Semiconductor techniques.
Multi-access
An arrangement whereby a computer can service several users apparently simultaneously.
Multi-programming
The running of several programs concurrently under the control of a Supervisor program.
Multi-processing
The use of two or more processors in one system to share the load.
Operating System or O.S.
A set of programs for the overall control of a computer system.
Off line
Operations carried out without direct computer control. After all the data has been set up by computer, and stored e.g. on magnetic tape, the task is carried out on peripheral equipment.
On line
Operations carried out under direct computer control.
Parallel structure
Data structure in the form of parallel lists.
Photochromic Tube
A Dark Trace Display. See page 14.
Plex structure
Complex data structure.
Polling
Interrogating of terminals in turn by C.P.U. (q.v.)
Random scan
C.R.T. scanning system whereby vectors and symbols are drawn without reference to the previous beam position.
Rand Tablet
A device for input of coordinate data. (See page 11.)
Raster scan
A television type of scan. (See page 10.)
Real time
Processing carried out on incoming data where the results are sent out in time to modify that data. E.g. in process control.
Re-entrant supervisor
Supervisor program which utilises pure procedures, i.e. it will handle the operation of a computer in a time-sharing mode.
Refresh rate
The rate at which a display has to be regenerated to remain steady. (See page 7.)

Register
   A working store in a computer or display, where data is held ready to be used.
Ring Structure
   A system of organising data in store. (See page 21.)
Rolling Ball
   A device for input of coordinate data. (See page 10.)
Rubber-band line
   Display of a straight line which has one end fixed and the other end following the light pen.
Satellite processor
   Small processor which is backed up by a larger machine.
Scaling
   Magnifying or reducing a picture on a display. (See page 38.)
Scan Converter
   A type of storage display. (See page 12.)
Scissoring
   Displaying a portion of a picture to fill the C.R.T. screen. (See page 37.)
Scope
   A C.R.T. complete with driving electronics.
Sketchpad
   The first notable interactive display system. (See page 1.)
Software
   The users' programs, and the facilities of a computing system provided by programs. (Cf. Hardware)
Storage tube
   A C.R.T. capable of holding a picture without refresh. (A DVST.) (See page 13.)
Supervisor
   Part of the Operating System which organises scheduling.
Teletype
   An electric typewriter with communication facilities to and from a computer.
Time-sharing
   Use of the computer by several users at once, achieved by means of an Executive program which allocates processing time to each in turn.
Touch Wires
   A system for the input of data by placing fingers on operating points on the screen. (See page 11.)
Tracking Cross
   A cross displayed as a reference point for pointing the light pen. (See page 33.)
Tree
   A branching sequence of possible operations, where decisions may be made at each branch as to which of two paths to follow.
Vector generation
   The drawing of lines on a display.
Williams Tube
   C.R.T. on the face of which information can be stored and subsequently re-read by scanning. The stored information is in the form of electric charge and is not visible.

Window
A section of a complete drawing shown on a display. (See page 37.)
Word length
The basic unit of information as operated on by the computer. Interpreted as data or programming instructions, the word is often 16 or 24 bits long, or of variable length made up of a certain number of bytes depending on the program requirements. (See Bit, Byte.)
Zooming
Continuous scaling and scissoring. (See page 38.)

Consolidated Bibliography

A list of references containing the computer graphics bibliography of each chapter plus further recommended published work immediately relevant to the study of the subject.

Applied Data Research Inc. "Autoflow".
Charp, S., and Wye, R. E. "An overview of Project Grow—A computer assisted instruction system for the School District of Philadelphia". To be published in *Educational Technology*.
Computerworld. "Computers at Hemisfair 1968," April 10, 1968, p. 8.
Coons, S. A. "Surfaces for computer-aided design of space forms", MAC-TR-41 MIT Project, June 1967.
Coons, S. A. "Surfaces for computer-aided design of space figures", Mac Project. Intern. Memo MAC-M-255, July 1965.
Culler, G. J. "Users Manual for an on-line system", *On-Line Computing*, pp. 303–324, 1967.
Davis, M. R., and Ellis, T. O. "The Rand Tablet: A man-machine graphical communication device", *AFIPS Conf. Proc.*, Vol. 26, FJCC 1964, pp. 325–331.
Dertouzos, M. L., and Graham, H. L. "A parametric display technique for on-line use", *Proc. SJCC*, 1967, pp. 201–209.
Dijkstra, E. W. Letter to the editor, *Comm. ACM*. 11, March 1967, pp. 147–148.
Donaldson, P. E. K. "Error decorrelation: A technique for matching a class of functions", *Proc. III. International Medical Elect.*, pp. 173–178.
Doran, J. E. "New developments of the graph traverser", *Machine Intelligence* 2, pp. 119–135.
Doran, J. E., and Michie, D. "Experiments with the graph traverser program", *Proc. R. Soc.* (A), **294,** pp. 235–259.
Electronic News. "Lockheed using computer design", April 1, 1968.
Electronics. February 19, 1968, pp. 50–51.
Ellis, T. O., and Sibley, W. L. "On the problems of directness in computer graphics", *Proc. Emerging Concepts in Computer Graphics*, Fall 6, **67,** pp. 123–130.
Friedlander, G. D. "Automation comes to the printing and publishing industry", *IEEE Spectrum*, April 1968, pp. 48–62.
Gruenberger, F. "Computer Graphics—utility—production—art", Academic Press, 1966.
Haibt, L. A. "A program to draw multi-level flow-charts", *Proc. Western Joint Comp. Conf.*, 1959, pp. 131–137.
Hain, G., and Hain, K. "Automatic flow chart design", *Proc. 20 Nat. ACM Conf.*, 1965, pp. 513–523.
Information Inc. "On-line computing systems", *Proc. Symp. Los Angeles, California*, 1965, p. 152.
Jacks, E. L. "Observations on a graphic console system", G.M.C. (Unpub.).
Jacks, E. L. "A laboratory for the study of graphical man-machine communication", *AFIPS, Conf. Proc.*, Vol. 26, *FJCC*, 1964, pp. 343–350.
Johnson, T. E. "Sketchpad III—A computer program for drawing in the dimensions", *AFIPS Conf. Proc.*, Vol. 27, *SJCC*, 1965, pp. 347–353.
Journal of the Society for Information Display.

Kapps, C. A. "Sprint: A direct approach to list processing languages", *SJCC Proc.*, 1967, pp. 677–683.

Keast, D. N. "A survey of graphic input devices", *Machine Design*, August 3, 1967, pp. 114–120.

Kennedy, J. R. "A system for time-sharing graphic consoles", *Proc. SJCC*, 1967, pp. 211–222.

Knowlton, K. "Computer—Animated movies", Bell Tel. Labs. Inc. January 1968. To be published *Proc. Conf. Emerging Concepts in Computer Graphics*, University of Illinois, 5–8, November 1967.

Knowlton, K. "A computer technique for producing animated movies", *AFIPS Conf. Proc.*, Vol. 25, *SJCC*, 1964, pp. 54–87.

Knuth, D. E. "Computer drawn flowcharts", *Comm. ACM.*, **6**, 9, September 1963, pp. 555–563.

Kulsrud, H. E. "A general purpose graphic language", *Comm. ACM.*, April, 1968, 8, pp. 247–254.

Larkin, F. M. "A combined graphical and iterative approach to the problems of finding zeros of functions in the complex plane", *Computer J.*, **7**, October 3, 1964, pp. 212–219.

Larkin, F. M. "The structure and implementation of ghost", Spec. Sess. Brunel.

Lewin, M. H. "An introduction to computer graphics terminals", *IEEE Proc.*, Vol. 55, No. 9, September 1967.

Licklider, J. C. R. "Man-machine symbiosis", *IRE Trans. on Human Factor in Electronics*, Vol. HFE-1, March 1960.

Licklider, J. C. R., and Clark, W. "On-line man machine communication", *Proc. SJCC, San Francisco, California*, May 1962, Vol. 21, pp. 113–128.

Loewe, R. T., et al. "Computer generated displays", *Proc. IEEE*, Vol. 49, No. 1, January 1961, pp. 185–195.

Machover, C. "Graphic CRT terminals—Characteristics of commercially available equipment", *AFIPS Conf. Proc.*, Vol. 31, 1967, Fall Joint Comp. Conf.

McGee, W. C., and Howry, S. K. "On-line SUMX", Report No. 320–3230, January 1968, IBM Palo Alto Scient. Center.

Michie, D., and Chambers, R. A. "Boxes: An experiment in adaptive control", *Machine Intelligence*, **2**, pp. 137–152.

Michie, D., Fleming, J. G., and Oldfield, J. V. "A comparison of heuristic, interactive and unaided methods of solving a shortest route problem", *Machine Intelligence*, **3**, Edin. Univ. Pr., pp. 245–255.

New York Times. "Baby robot learns to navigate in a cluttered room", April 10, 1968, p. 49.

Ninke, W. H. "Graphic 1—A keynote graphical display console system", *Proc. FJCC*, 1965, pp. 839–846.

Noll, A. M. "Stereographic projections by digital computer", *Computers & Automation*, May 14, 1965.

Ophir, D., Rankowitz, S., Shepherd, B. J., and Spinrad, R. J. "Multi-console computer display system", Brookhaven National Lab. Report 11589, 1967.

Orr, W. D. *Conversational Computers*, J. Wiley & Sons Inc., 1968.

Parker, D. B. "Solving design problems in graphical dialogue", Ch. VII—*On Line Computing*, Ed. Karplus, W. J., McGraw Hill Co., 1967.

Poole, H. H. *Fundamentals of Display Systems*, Spartan Books, Washington.

Richardson, F. K., Lo, T. Y., and Gear, C. W. "Computer aided programing system", *Proc. Emerging Concepts in Computer Graphics*, Fall 67, pp. 171–184.

Rippy, D. E., and Humphries, D. E. "Magic—A machine for automated graphics interface to a computer", *AFIPS Con. Proc.*, Vol. 27, Part 1, *FJCC* 65, p. 819–830.

Roberts, L. G. *Graphical Communications and Control Languages*, Second Congress on the Information System Scienc. Spartan Books, 1964.

Roberts, L. G. *Machine Perception of Three Dimensional Solids*, Rep. 315, Lincoln Libr., MIT., May 22, 1963.

Rose, G. A. "Light pen facilities for direct view storage tubes—An economical solution for multiple man-machine communication", *IEEE Trans. on EC*, Vol. EC12, No. 4, August 1965, pp. 637–639.

Ross, D. T., and Rodriguez, J. E. "Theoretical foundation for the computer aided design system", *AFIPS Con. Proc.*, Vol. 23, SJCC, 1963, pp. 305–322.

Schwartz, J. I., et al. "A general purpose time-sharing system", *AFIPS Conf. Proc.*, Vol. 25, SJCC, 1964, pp. 397–411.

Sherman, P. M. "Flowtract—A computer program for flowcharting programs", *Comm. ACM.*, **9**, 12, December 1966, pp. 845–854.

Siders, R. A. "Computer-aided design", *IEEE Spectrum*, November 1967, pp. 84–92.

Stotz, R. "Man-Machine console facilities for computer-aided design", *AFIPS Proc.*, Spring Joint Comp. Conf., 1963, p. 323.

Sutherland, I. E. "Sketchpad, a Man-Machine communication system", *AFIP Proc.*, Spring Joint Comp. Conf., 1963, p. 329.

Sutherland, W. R. *On Line Graphical Specification of Computer Procedures*, Tech. Report. *ESD-TR*-66-211, *MIT* Lincoln Lab., May 1965.

U.S.A. Department Agriculture Publication. *Technology and Man-Power in Design and Drafting*, 1965 75.

Vorhaus, A. H. "General purpose display system", *Datamation*, July 1966, pp. 59–64.

Weisberg, D. E. "Man-Machine communication and process control", *Data Process Magazine*, September 1967, pp. 18–24.

Wood, L. H., Ely, C. A., Glanzer H., and Radice, V. "The Amtram input-output terminal", *Computer Design*, March 1968, pp. 68–74.

Index

Accuracy 39
ACM 61
Adage 74, 75
AED Project 213
ALGOL 31, 34, 88, 111, 212, 213
Alphanumeric Inc. 71
Alphanumerics 4, 85, 102
American Standards Association 49
AMTRAM 78
Animation 17
APACE Centre, Aldermaston 87
Applications Programs 25
Applications 2, 3, 6, 62 ff., 68, 73, 85, 87, 102, 104, 106, 119, 172, 187 ff.
 Aero Engines 161
 Aircraft Engineering 62, 86, 87, 161, 199
 Air Defence 61
 Airline Reservations 48, 58, 86, 105, 195
 Air Traffic Control 46, 86, 189
 Architecture 87, 125 ff., 201
 Automobile Engineering 63, 196, 201, 210
 Business Management 3, 57, 65, 85, 106, 119 ff., 198
 CEGB 86
 Chemical Engineering 38, 68, 88
 Closed loop 69
 Continuous Path Programming 62
 Disc Design 163
 Education 69
 Electricity Undertakings 17, 18, 86
 Electronics Design 63, 87, 169, 173, 174, 198
 Filters, Electronic 173
 Financial Analysis 3
 "Firtree Root" Design 164
 Gas Undertakings 86
 Graphic Arts 71
 High Energy Physics 149 ff.
 Hospitals 58, 85, 104, 106, 204
 Integrated Circuit Design 63
 Libraries 85
 Machine Intelligence 177
 Mechanical Design 161
 Newspapers 71, 72
 Nuclear Research 88, 135 ff., 149 ff.
 Pattern Recognition 71
 Pharmaceutical 68
 Printed Circuit Boards 174
 Printing 71, 105
 Process Control 68
 Satellite Program 2, 68
 Shipbuilding 208
 Simulation 66
 Steel Mills 86
 Stock Control 105, 119 ff., 199, 200
 Structural Steelwork 87
 Textile Design 64
 Topography 193

Applications—*continued*
 Urban Planning 3
 Wind Tunnels 68
APT 19
ARDS 106, 209, 210
Argonne National Lab. 198
ARGUS 223
ARO 68
ARQ 49
ASCII 95, 100, 120, 192
ATLAS 181
Atomic Energy Commission 109
Autoflow 111, 198
AUTOGRAPH 171
Autonomous Access 17

Background data 57
Back up equipment 9
Beckman Instruments 203
Bell Telephone Laboratories 64, 72, 79, 105, 106, 208, 210
Benson–Lehner 79, 137
Bernstein, M. I. 187, 192, 197
Bird, S. 17, 27, 189, 190, 213
Birmingham University 88
Blackall, P. M. 149, 158
Blink 50, 55
"Blobs" 126–33
BOADICEA 86, 195
Boeing Airplane Co. 62, 65, 72
Bolt, Beranek, Newman 67, 74, 78, 81
Borger, R. 203
BOXES 181–4, 203
BRAD 105
Brainerd, Prof. John 72
Brentwood School, Calif. 69
British Steel Corp. 87
Brookhaven National Labs. 105
Brown University, R.I. 4, 73
Brunel University (Brunel Symposium) 106, 107, 187, 207, 210, 217
Bubble chamber measurements 151
Bunker-Ramo Corp. 74, 75
Business applications. *See* "Applications—business management"
Button box 31, 66, 161

Calcomp Ltd. 219
California Computer 79
California, University of 73
Cambridge University 87, 88, 214
Carter, A. W. H. 199
Category Selection 45
CATV 91
C.B.S. Laboratories 71
CCI 103
CERN 149, 158, 159
Chadwick, E. G. 192, 202
Chambers, R. A. 177, 181, 185, 203
Character & Vector Generation Techniques 9, 50, 53, 96, 192

Index

Charactron 9, 12
Cheek, T. B. 209
Circle Generation 56, 189
Clifton, J. S. 206
COBOL 3
Colour Display 79, 103
Columbia University 67, 71, 73
Common Language 2
Computer Control Company 79
Computer Displays Inc. 74, 78, 106. 157, 196, 206
Computer Instrumentation Ltd. 220
Conducting glass overlays 11
Conductron Inc. 81
Congress of the International Federation of Information Processing 4
Construction, materials and design 126
Control Data Corp. (CDC) 74, 75, 79, 149, 155, 157
Cossor Electronics Ltd. 103, 119, 221
Costs 91 ff., 100, 101, 102, 106, 113, 120, 125, 202, 207 ff., 209, 210, 211
C.R.T. Displays, principles 7, 41, 57, 61, 91, 92, 93, 99, 104, 211
C.R.T. Terminals, manufacturers of 73, 103, 217 ff.
Culler–Fried On-Line Console 155

DAC System 63
Daniels, T. 188
Daresbury Nuclear Physics Labs. 88
Dark Trace Display 13
Data Reduction 2
Data Structure 4, 19, 191, 212. 213
Data Tablet 81
Data Transfer 48 ff.
Data Transmission Techniques 100, 122
Datel 600 Service 100
Decision Loop 181
Deflection Systems 12
Delay Lines 42
Deletion 25, 103
Design, Computer-Aided 62 ff., 95, 125 ff., 161 ff., 173, 210
Design principles, hardware 7
Detection of computer fault 145
Digital Equipment Corp. (DEC) 74, 75, 78, 79, 97, 106, 113, 114, 180, 208, 222
Digitising, problems 194
DISMAN 88, 212
DISPLAY 66
Display Commands 17, 130, 212
Display file 17, 209
Displays, graphical 3, 4, 11, 17, 29, 41 ff., 63, 64, 67, 69, 70, 71, 77, 85 ff., 91 ff., 99 ff., 129, 135, 139, 149 ff., 173, 174 ff., 177, 182, 183, 192, 206, 207, 210, 212, 215, 221 ff., 229, 230
Djikstra, E. W. 112
Doran, J. E. 180
Dot Matrix 10

Dot Vector 10, 56
Durrands, K. J. 196
DVST 13, 93, 94, 95, 96, 195, 196, 211

ECAP 172, 202
EDGAR 88
Edinburgh University 87, 185, 208
Editing 103
Einstein, Albert, College of Medicine 71, 73
Electron Beam Recording 228
Electronic Mixing 57
Elliot Green, R. 99, 107, 207
Elliott-Automation 103, 181, 184, 212
Elliott Computing Services 39
Errors. (*See* "Mistakes")
Evans, D. R. 7, 14, 188
Expansion 19, 20, 21

Fairchild Semi-Conductor 64, 79
Ferranti Ltd. 86, 97, 223
Ferrodot 105
Fitz, A. E. P. 119, 123, 198
Flexibility, display 55, 207 ff.
Flicker, problem of 195
Flowcharts 111 ff.
Ford Motor Company 63, 196
Formatting 53
FORTRAN 2, 3, 31, 88, 111, 112, 157, 162, 198, 201, 212 ff.
Franklin, D. A. 187, 204, 205
FRED 88
Friden 79
Future Developments 5, 12, 96, 105, 116, 156, 158, 207 ff.

Gas Discharge Matrix 14
Gas Discharge Tubes 188
Gear, Prof. C. W. 109, 117, 189, 197
GEMS 171
General Electric 64, 80
General Motors 62, 63, 81, 196
Geometry 165, 166
GEO Space Corp. 79
George, I. 189
Gerber Drafting Machine 169
GHOST 137, 138, 212
Gibson, Dr. A. 139, 145
GLANCE 105, 210
GPAK 201
GPO 100
GRAIL Project 109, 113
Graphs, display of 192
GRAPHIC-2 208
Graphic Displays Ltd. 106, 224
Graphics, importance of 2
Graphics, statistics of 61, 85
Gravity Fields 34, 35, 36
Grid Technique 35
GRIN-2 208
Guedj, A. 188

Haibt, L. A. 112
Hain, G., and Hain, K. 112

Index

Hard Copy 103
Hardware Techniques 7, 137, 157, 169, 170, 187
Harris Intertype 71
Harvard University 71, 73
Hash Coding 4
Hazeltine 79
"Hemisfair 68" 64
Herzog, Prof. B. 194, 196, 205
Hewlett Packard 79
Holography 81
Hopkins, John, University 72
Huggin, Prof. W. H. 72
Humphreys & Glasgow Ltd. 88
Hutchinson, D. S. 198

IBM 1, 6, 64, 74, 75, 87, 113, 129, 157, 162, 169, 171, 194, 198, 201, 202, 205, 208, 213, 225
ICL 29, 39, 88, 212, 226
IDI 65, 74, 75, 79, 83
IEEE 61
ILLIAC 113, 114, 194
Illinois Institute of Technology 73
Illinois University 73, 109, 113, 114, 116, 194
Imperial College 87, 88, 208, 213
Information International Inc. 74, 194
Input/Output, Techniques, Principles 4, 10, 51 ff., 80, 81, 116, 132, 135, 157, 158, 162, 169, 175, 194, 207 ff., 217
Institute of Computer Science 208
Intelligent Machine 179, 203
Integrated Circuits, Design 63
Integrated Systems 75
International Telephone & Telegraph Corp. 74
INTRACS 206

Jacks, E. L. 63
Jacquard Loom 1
Joint Computer Conference, AFIPS 1
Jones, J. R. A. 198
Joystick 10, 96, 193

Keyboards 10, 31, 66, 129, 155, 161
Kilgour, A. C. 208
King's College Hospital 204
Kirstein, P. T. 208
Knuth, D. E. 111
Kulsrud, H. E. 81

Lang, C. A. 214
Languages, Computer 2, 19, 81, 88, 111, 112, 114, 157, 158, 195, 198, 207
Larkin, F. M. 135, 137, 140, 142, 191, 201, 212
Lassalle, J. C. 149, 158
Lawrence Radiation Laboratory 72
Lee, B. J. 211
Leventhal, Prof. C. 67

Light pen 11, 31, 32, 33, 34, 64, 96, 103, 129, 151, 152, 158, 161, 172, 180, 182, 183, 191, 193, 210, 214
Linders, J. G. 208
Lines, display of 34
Lissajous Figure Systems 10
Lo, T-Y. 109
Local Memory 8
Lockheed Georgia 62, 87
Los Alamos Scientific Labs. 72
Ludley, J. H. 190

Macdonald, R. I. 195
McDonnel-Douglas 62
McDouall, D. C. 169, 177
Machine Intelligence 203
Machover, C. 61, 83, 193
Magnetic Well 142
Manipulation 25
Man-Machine Communication 3, 5, 11, 17, 87, 129 ff., 177, 179 ff., 203
Marconi Company Ltd., The 27, 88, 190, 213, 227
Massachusetts Institute of Technology (M.I.T.) 1, 67, 73, 78, 106, 109, 206, 209, 213
Mathematical Analysis, On-line 155
Matricon 10, 13
Matrix Displays 14
Matsa, S. M. 1, 6, 187
Menu 32, 130
Mergenthaler 71
Meyer, Dr. E. 67
Michie, Prof. D. 179, 180, 181, 185
Michigan, University 73
Microfilm Images 228
Ministry of Defence 119
Ministry of Technology 171, 196
Minnesota, University 71
Mistakes, Problem of 31, 32, 49, 112, 145, 201
Modelling Data 3
Monitor Systems 79
Monoscope 9
M.O.S. 95, 96, 106, 210
Motorola 64
Movies, Computer Generated 72
Multi-person Access 196
Multiple Display Terminals 54 ff. 99 ff., 207 ff.

N.A.S.A. Space Flight Center 68, 78
National Engineering Laboratory 86, 87, 208, 212
N.C.C. (National Computing Centre) 85, 89, 196
N.C.E.E.F. (National Committee for Electrical Engineering Films) 72
Newman, W. M. 87, 208, 213, 214
New York Schools 69
New York University 73
Nicholson, A. W. 187
Nightingale, A. F. 202
Ninke, W. H. 210

Index

Ohio, University 73
Oldfield, Dr. J. V. 87
Operating Systems 26
Orbits, charged particle 143
Orthogonal Projection 39
Output. (*See* "Output/Input")
Overlays 57
Oxford University 88

Parish, C. C. M. 195
Parslow, R. D. 204
Passive Enquiries 119, 122
P.D.P. (*See* "Digital Equip. Corp.")
Pennsylvania, University 72, 73
Peters, B. 125
Phase Fluid Behaviour 145
Philadelphia Schools 69
Philco–Ford 69, 74, 79
Phosphors 12
Photochromic Glass 96
PICPRINT 171
Picture Scan Techniques 10
Pinson, E. N. 208
Pitteway, Prof. M. L. V. 209
Plessey 103
PL/1 2, 3, 31, 111, 169
Plotters 79, 101, 210, 219, 220, 224
Poole, H. H. 41, 59, 192
Price, A. G. 189, 201
Pricing. (*See* "Costs")
Programming Systems 2, 109, 163, 189, 198, 201
Program Structure 20, 22, 29, 30, 31, 170, 190
Prowse, R. W. 187
Purcell, P. A. 201

Racal Electronics 87
Rafferty, J. S. 212
Rand Corporation 109
Random Position Monitors 79
Rand Tablet 11, 81
Raytheon Company 59
R.C.A. 64, 69, 71, 97
Reaction Handler System 208, 213, 215
Reading University 211
Rear Projection 57
Recording System 228
Reed College 73
Refresh 7, 42 ff., 54, 91, 92, 93
Remote Communication 99
Research, Computer Graphics 207 ff.
Retrieval of Data 3, 4, 95
Rewrite Time 96
Richardson, F. K. 109
Ring Structure 21, 190, 212
Roberts, C. T. 202
Roberts, Dr. K. V. 139
Robot Control 71
Rolling Ball 10, 193
Rolls Royce Ltd. 87, 161, 167, 201, 212
Rosing, R. E. 193
Ross, D. T. 213
Rotation 4, 50, 52, 56

Royal Radar Establishment 14
Ruben, M. A. 91, 97, 195, 209
Rundle, A. R. 29, 39, 190, 212
Rutherford Laboratories, Harwell 208

SAGE System 61
Sanders Associates 74, 103
Sandford-Johnson, P. B. 203
Satellite Computers 100, 157, 188
Saturn IV Rocket 68
Scaling 38
Scan Conversion 12
Scanning Techniques 9, 91, 92, 93, 212
Scissoring 37, 38
Science Research Council 185
Scientific Control Systems Ltd. 107, 207
Scientific Data Systems Inc. 74
Selective Erasure 13
Shaped Beam 9, 11
Sherman, P. M. 111
Shoe-Box 205
Siders, R. A. 74
SIGMA 171
SKETCHPAD 1, 39, 61, 66, 113
Society for Information Display 61
Software Techniques 17 ff., 29 ff., 81, 88, 120, 138, 157, 162, 170, 187 ff.
Spark Chamber Measurements 152
SPRINT 114
Standardisation 88
Stanford Research Institute 71
Stanford University 73
STC (Standard Telephones & Cables) 87, 177, 203
STL 105
Storage Techniques 4, 11, 12, 13, 42, 78
Stotz, R. H. 196, 206, 209
Stroke Writing Systems 11
Stromberg–Carlson Corp. 74, 79, 169
Subpictures 25
Superimposition 57, 58
Sussex University 88
Sutherland, Dr. I. 1, 39, 61, 113
Sutherland, W. R. 109, 113, 116
Sylvania 80, 81
Symbols 31
Syracuse University 72
System Development Corp. 65, 66, 67, 197
Systems Engineering Labs. Inc. 74, 75
System Expansion 52

Tabular Displays 11, 53
Tasker Instruments Corp. 74, 81
Taylor, Dr. F. E. 85, 99, 195, 205
Tektronix 78, 157, 195, 209, 229
Teledeltos Paper 211
Teleputer System 78
Telephone Connections 99, 100, 172, 208
Teletype 99, 101, 121
Texas A & M 67
Texas Instrument 64

Thomas, Dr. R. E. 193
Thomas Electronics 80
Three-Dimensional Graphics 39, 81, 214
3M Company 228
Time Sharing 1, 188
Torson, B. T. 161, 167, 201, 212
Touch Wires 11
Tracking 33, 34, 191
Tradeoffs 5 ff.
T.V. Systems 79

UAIDE 61
UCLA 71, 73
UK 85 ff.
UKAEA Culham Laboratory 135, 136, 212
United Aircraft 64
UNIVAC 65, 74, 119, 208, 213, 230
U.S.A. 61 ff.
Utah, University 73, 213

Van Dam, Prof. A. 4
Van der Eycker, G. G. J. A. 193
Vandoni, C. 149, 159
Vectors 9, 10, 50, 53, 56
Videocomp 71

Walter, P. E. 125, 133, 201
Weiner, Prof. D. D. 72
Weisberg, D. E. 68
Westinghouse 65
West Sussex County Council 87, 125
Williams Tube 211
Windowing 19, 20, 21, 37
Wisconsin University 71
Wood, B. L. 195
Word Formats 48
Wycherley, R. D. H. 212

Yule, A. P. 149, 159

Zooming 38

97801